ONE TRACK MIND

a trainspotter's journey

ONE TRACK MIND

a trainspotter's journey

Written by

NICK CARTER

Published by *begin-a-book* Independent Publishers
www.beginabook.com

Copyright © 2025 by Nick Carter

All rights reserved.

No portion of this book may be reproduced in any form without written permission from the publisher or author, except as permitted by U.K. copyright law.

This publication is designed to provide accurate and authoritative information in regard to the subject matter covered. It is sold with the understanding that neither the author nor the publisher is engaged in rendering legal, investment, accounting or other professional services. While the publisher and author have used their best efforts in preparing this book, they make no representations or warranties with respect to the accuracy or completeness of the contents of this book and specifically disclaim any implied warranties of merchantability or fitness for a particular purpose. No warranty may be created or extended by sales representatives or written sales materials. The advice and strategies contained herein may not be suitable for your situation. You should consult with a professional when appropriate. Neither the publisher nor the author shall be liable for any loss of profit or any other commercial damages, including but not limited to special, incidental, consequential, personal, or other damages.

Book Cover by AnnMarie Reynolds for *begin a book* Independent Publishers

Photography © by Nick Carter - including cover photography

Print Edition First published in the United Kingdom in June 2025

ISBN Print - First Edition 978-1-915353-30-6

Published by begin-a-book Independent Publishers
www.beginabook.com

For Tracey, George, Matt, Ben and Brice.

My wonderful family.

A Note from the Author

Dear Reader,

Thank you so much for buying this copy of 'One Track Mind'. I hope you find something within its pages that resonates with your interest in the hobby.

In an earlier draft, many of the pages included loco numbers seen on the various trips referenced, however I received some helpful feedback suggesting that the addition of the numbers broke up the text too much. Therefore, with a few exceptions, I have deleted the loco numbers from the body text.

To compensate, I have pulled together a simple spreadsheet that includes, in date order, many of the locos seen for reference purposes. I should add that this list does not attempt to be an authoritative guide to the many instances where locos have carried more than one number, although I have added a few comments against some numbers. You can download a copy of this PDF using the QR code at the bottom of this page or alternatively you can visit *https://tinyurl.com/5cjf33zw*.

This book has been in the making for a few years and I enjoyed going back over my old notebooks as I worked through the chapters, although this often provided a great distraction which would take me away from the task in hand! I would love to hear from readers with any comments on the book or from those who just want to share their own stories.

Please feel free to contact me on *nick.carter61@outlook.com*.

Thank you for reading, and I hope you enjoy my journey as much as I did - and will continue to do so.

Nick.

CONTENTS

Glossary	8
Acknowledgements	11
Introduction	13
Early Days 1971 - 1975	17
Bristol Bath Road Depot	33
Branching Out	43
Summer Saturdays At Temple Meads	47
London	61
More Trips And Open Days	69
The Big Bike Ride ~ June 1981	87
Freedom Of Scotland Trip	117
Towards The End Of The Century	129
Rolling Into The Noughties	143
The Future Is Bright	159
Bibliography	167

Glossary

(includes most common references found in the following pages)

ABC Combined Volume	Book produced by Ian Allen (later replaced by Platform Five Publishing) detailing the numbers of all the locomotives and other rolling stock. There were individual booklets for engines, multiple units, carriages and also a 'combined' volume which pulled them all into one book.
APT	Advanced Passenger Train
Baby Deltic	Nickname for a class 23, the engines were half the size of the larger Deltics, hence the 'Baby' name. Initially very unreliable but after modification became more efficient, however they were considered non-standard as a result of the 1960s National Traction Plan which streamlined the locomotive fleet.
Bunking	Going around a depot without permission.
Chopper	Nickname for a class 20 reflecting the sound of the engine resembling a helicopter.
Cop	When you see an engine for the first time.
Deltic	Nickname for a class 55 due to the arrangement of the engine pistons which form a delta shape.
DMU	Diesel Multiple Unit - one or more carriage train powered by on-board diesel engines

DRS	Direct Rail Services - freight operator, best known for operating nuclear flask trains.
Duff	Nickname for a class 47 reflecting their failure rate when first introduced.
EMU	Electric Multiple Unit – One or more carriage train using on-board electric motors powered by a third rail or overhead power cable.
EWS	English, Welsh and Scottish - major freight operator.
Gronk	Nickname for a class 08 Shunter.
Growler	Nickname for a class 37 due to the sound of their engines.
HST	High Speed Train
Hymek	Nickname for class 35 due to their Meydro hydraulic transmission system.
Locoshed Book	This pocket-sized booklet listed all the engine numbers and, importantly, to which shed they were allocated.
Locoshed Directory	This invaluable Ian Allen booklet listed all the engine sheds and directions to reach them. A must-have resource for the railway enthusiast.
Peak	Nickname for class 44, 45 and 46 diesels. The first ten (class 44) were named after mountains in England, Scotland and Wales.
Rail Express Systems (RES)	This was a business sector of British Rail that was responsible for the transport of mail and parcels, however much of this traffic had moved from rail to road by the early 2000s.
Shed Bash	A trip where a number of sheds are visited.
Shed	Nickname for a class 66 due to the apex design of the front cab.
Slim Jim	Nickname for a class 33/2 due to their narrower bodies compared with the standard class 33, to enable them to work through the narrower tunnels between Tunbridge Wells and Hastings. Class 33/1s were also known as 'bag pipes' due to the array of pipes on the front of the engines to enable multiple push/pull working with EMUs. Whole class also known as 'Cromptons' as they were fitted with Crompton Parkinson Generators.
Toffee Apple	Nickname for class 31 sub class as the shape of the power controller resembled the shape of a toffee apple. Other nicknames were 'skinheads', for those locos without a head code box and 'peds' reflecting their somewhat pedestrian performance.
TOPS	Total Operating Processing System - computer system used by British Rail for managing locomotives and rolling stock.

Acknowledgements

Firstly, I would like to thank my very patient wife, Tracey, for putting up with me going on about railways for so long. An equal appreciation of thanks also goes to our offspring, George, Matt, Ben and Brice for having to suffer a similar fate.

I would also like to thank AnnMarie Reynolds from *begin-a-book Publishing* who has been so supportive in helping to make this book see the light of day. It would not have happened without her and I recommend AnnMarie to anyone who has an itch to get a book out there.

I should, of course, also mention my lifelong friend Brian, who features throughout the book.

I would also like to thank some of my fellow enthusiasts on Twitter / X who provided such useful feedback on an earlier draft. In particular I would like to thank Mark Fletcher, David Hudson, Andrew Liszka and Keith Byrne.

Introduction

I should say from the outset that I have never owned the one item that most people would say is synonymous with trainspotters, namely the humble duffle bag, used by many to carry their notebooks, a sandwich and a drink. I have, however, owned that other indispensable item - an anorak, (as most of us have if we go by the dictionary definition of a 'jacket to keep the rain off'). The trainspotter is often portrayed as a figure of fun, grouped together on the ends of platforms scribbling down numbers, oblivious to the rest of the world, but for me and many others, it was so much more than that.

I consider myself lucky as I have many interests, but my fascination with railways has been around the longest, popping in and out of my life as the years rolled by. I'm not good at remembering dates of birthdays and anniversaries but ask me when the last class 52 Western diesel hauled special ran (26 February 1977) or the date when I first went around York shed (7 April 1973), and I can reel them off to order. From the age of ten onwards, railways have been in my life. On a bookshelf at home, there is a row of A5-size ruled notebooks going back to 1971, each page is ruled into four columns and contains lists of dates, locations and numbers. I can pick up any of these little books and find keys to open all sorts of memories. Sometimes my preoccupation with railways has been the most important thing in my life, whilst at other times, it has lurked somewhere in the background, crowded out by other events but never quite out of sight.

I am by nature a bit of a hoarder and have kept almost all my old railway books and miscellanea. This means that regardless of my level of preoccupation and interest, I've always got my loyal hobby to come back to. I'm not sure how useful a hobby it is though. I'm good at reading train timetables, I have a sound knowledge of the geography of England, Wales and Scotland, I understand the 24-hour clock, and I'm quite handy at finding my way around most cities and following street maps, but what else? Well, it's provided many hours of enjoyment, fantastic memories and has been a constant through life when times have been challenging. Conversely, I realise that despite having visited most cities on the British Rail network, I know little about them other than the location of the station and loco depot.

There are some where I know the football ground too – the great game being my second love - and to be more precise, Bristol Rovers Football Club, known affectionately as 'The Gas' to its followers (the nickname hails back to when they played at Eastville next to a gas works which resulted in a related smell often being in the air).

I also wonder how many other things I could have done if I had not spent most of my teenage years loitering around railway stations. I could have spent more time trying to play the guitar and perhaps become a lead guitarist with one of the fledgling punk bands that were springing up in the mid-1970s. I may even have got exceptional exam results rather than the average ones that I did pick up, and who knows what might have been? Oxbridge and a glittering career in a high-flying profession?

Whatever may or may not have happened is conjecture. The reality is that much of the 70s and 80s were spent travelling all over the UK rail network and paying sometimes daily visits to Bristol Temple Meads station. If I was not spotting or photographing trains, I was reading about them or planning marathon trips around the country visiting engine sheds.

Every month I bought the Railway Magazine, which was invaluable at the time as it was the only means by which you could find out when engines had been reallocated to different loco depots - although inevitably, by the time you received the information, information, it was already a few months old. Nevertheless, each monthly edition resulted in an updating of the indispensable Ian Allen Locoshed book, which recorded every engine along with the shed where it was based.

One of the ironies today is that while the hobby is less actively pursued than in the 70s and 80s, there are more railway magazines available in W.H.Smiths (and other newsagents) now than at any other time. The whole spectrum of the hobby is covered; steam, the modern scene, railway modelling, the European railway scene, and even a magazine which specialises in the diesel period between steam and the current era. This one provides a wonderful flashback to my golden period.

Trainspotting, as a hobby, was in its heyday in the 1950's. At that time, any self-respecting boy (and I am not being sexist here as it has always been a predominantly male pastime) would be able to tell the difference between a Great Western King and Castle class steam engine. Since then, it is fair to say that its popularity has been in slow decline – which makes the current glut of available magazines a little surprising. It may be that these publications are mainly catering for armchair enthusiasts nostalgic for a bygone age. Their memories being awakened by photos of class 52 Westerns at Paddington or a pair of class 50s hammering up the infamous Shap bank with the Royal Scot. I have a daughter and son, George and Matt, and neither showed any interest in railways. Both are now in their thirties and following their own careers, untouched by the hobby that obsessed their dad - although I'm pleased to say that both are Rovers fans (Gasheads), so I did get something right. Neither of them even understands the concept of being fascinated by trains, so my interest is seen by the family as being a bit weird. My habit of visiting railway websites is similarly met with a sense of bewilderment, and my perusal of railway magazines in the newsagents is a further source of mirth.

As I have got older, my interest hasn't waned though the days of chasing locos around the country are long gone. I have discovered that the enthusiasm with which most hobbies are pursued, especially if they are taken up at a young age, becomes diluted when you encounter other responsibilities such as jobs, mortgages and raising a family, and this is very much what has happened with myself and railways. Furthermore, other areas of interest have crept into my life and taken up more of the time that was once reserved for my favourite pastime. Having said that, old habits die hard, and I find it impossible to drive over a railway bridge without being drawn to those parallel steel lines that trail off into the distance. I still love travelling by train, and through work I was able to tour around the network on a regular basis, occasionally taking the opportunity to visit an old railway haunt and see how much it has changed.

I should mention that two people appear throughout this book, my dad and my lifelong friend Brian. I think dads and granddads deserve a special mention. They are the unsung heroes who introduced so many of us to a love of trains (in many cases, they were probably reliving their youth), but we should not underestimate the impact they have had. I remember Dad telling me when he was a teenager during WW2 how he went by train on his own up to London from Bristol for the day. He went to Kings Cross and saw the big express engines on the buffers. Unfortunately, he didn't tell his mum of these plans, though, so received a bit of a telling-off on his return home!

I became friends with Brian at Sea Mills Primary School in Bristol. We lived about half a mile apart. I did not know at the time, as we battled to be milk monitors in Mrs Conway's class, that it would be a friendship which would span a lifetime. I'm not sure who influenced who, but we shared many interests: a love for railways, Bristol Rovers, rock music and, to a lesser degree, motorcycles. Although we now live in different parts of the country, we still get together to watch Rovers a few times a season.

Throughout the book I have inserted photographs that I have taken over the years, however I should mention that some of them are of dubious quality as a result of limited technical ability and poor quality equipment, however they are my photos and, like my notebooks, provide windows into my past and a simpler time. I should also add that I have avoided putting too many loco numbers into the text, however there is a link to a spreadsheet in the Authors Note at the beginning of the book which provides details of many of the locos seen on the trips that I have referred to.

I hope, through the following chapters, that you will enjoy some shared moments from my railway journey and that you will awaken your own memories and personal recollections as we travel together along those iconic tracks.

Early Days 1971 - 1975

I blame my Auntie Flo for the whole thing. Up until my tenth birthday, life had been quiet and predictable. We lived in Bristol in a tranquil council estate called Coombe Dingle in the north-west suburbs. Built just after the Second World War, our house was a semi-detached property in the shadow of Kings Weston Woods, part of the Blaise Castle estate. Council housing was the first and only choice for most people at that time, and in our street, we had a mix of neighbours, working class, middle class and a range of professions from dockers through to civil servants.

It was a good place to grow up, with the nearby woods providing childhood excitement. Dad had been in the Royal Navy for many years but left just before the birth of my sister, Ann, in 1957. I came along four years later. Dad then worked for the city council, chasing up people to pay their rates (Council Tax now), and Mum ran the home. I had been quite happily breezing through childhood pursuing the usual hobbies; playing with toy soldiers, making Airfix model plane kits, finding out about dinosaurs and doing other typical boy things. However, in March 1970, the real world came into sharp focus when a boy in my sister's class at Sea Mills Primary School was murdered on the nearby Shirehampton golf course. Police officers knocked on our door, making enquiries, but they never did solve the crime, leaving us to fear that a murderer was amongst us during our most vulnerable years.

The fact I can remember this clearly shows how much of an impact it must have had on me as a nine-year-old boy. Fortunately, I grew up in a loving family environment, so was provided with the security to help deal with such times.
1970s Britain was also the time when most public services were owned by the state. British Rail, National Express, British Telecom, British Airways, British Transport Hotels and British Gas ran the show, and if you went to the continent, it was likely to be on a British Rail Sea Link ferry.

The National Coal Board provided the coal for our fires, and the Central Electricity Generating Board provided our electricity. Locally, Bristol airport and the docks were owned by the city council, as were the buses which, in 1963, found themselves the subject of the Bristol bus boycott. This was a protest by black workers who were barred from working on Bristol's buses and eventually forced a (necessary) change in recruitment policy. It was later seen as a key event for race relations in the United Kingdom. For me back then, though, the highlight of the week was when the Alpine lemonade lorry delivered its bottles of multi-flavoured fizz - guaranteed to change the colour of your teeth and keep you eagle-eyed all night!

I digress. Getting back to Auntie Flo, it was April 1971, and mum thought it would be a good idea for the family to go to London, see a few sights and visit my Auntie who had left Bristol many years earlier. She'd married Uncle Howard, who died sometime before I was on the scene. He had a sign-writing business near Marble Arch so I assume that's how Auntie Flo came to live in London. The business eventually passed to their son, whose studio I once visited and I was instantly captivated by the intricate brushwork and intensity of the colours used, a skill which all but disappeared once computers arrived a few decades later.

Auntie Flo lived in a large rambling town house in Dollis Park, North Finchley. She was well into her seventies, smoked like a chimney and, whilst not very mobile, had the outlook and spirit of someone half her age.

As we did not have a car, Dad decided that we should go by train to visit her, and, as we never had much money, this was going to be a bit of a treat!

The trip started okay; my sister and I were generally well-behaved and had not yet become a pain to the other passengers – Dad, though, was worried. He was concerned that we would become bored, so, as a tactic to ward this off, he started to talk about railway engines and the different types of locos he had seen on his trips around the country travelling between various Naval bases. Although he was now working for the council, Dad remained in the Royal Naval Reserve and, consequently, for two weeks every year, would disappear to do his bit for Queen and country. Sometimes he would spend those two weeks on a shore base in the UK, whilst on other occasions, he would be flown out to some foreign destination. He told us that some of the locos he saw on his travels were named after warships, whilst others carried names such as 'Western Gladiator'.

We were pulling into the next station, Chippenham when a smug expression crossed Dad's face. I looked out and saw, at the head of a Bristol-bound express train, an engine named 'Western Gladiator'. It was as if Dad had it all planned! I was impressed, to say the least. This was stage management of a standard many dads could only dream about. The rest of the journey now took on a new twist as I looked out for engine names. Some I remember were 'Western Sultan', 'Sharpshooter' and 'Hercules', all of which I saw before we reached Paddington. The seed was definitely sown.

Although there was no further mention of trains during the week we spent with Auntie Flo, the return journey to Bristol, which happened to coincide with my tenth birthday, produced another six Westerns and one Warship-class diesel-hydraulic. I was ecstatic.

That being said, it was a slow germination period before the bug started to really take hold. Dad and I went on a couple of visits to Bristol Temple Meads station over the following months (usually when Mum wanted to go shopping in Broadmead), which left the two of us able to wander off to Temple Meads. The station was always busy and there was the added bonus of being able to see engines moving around Bristol Bath Road diesel depot, located alongside the station. Anything was better than going around the shops!

One of my earliest photos using an old camera that Dad bought in Hong Kong when he was stationed there with the Royal Navy. It's a shot of D1023 Western Fusilier standing at Platform 12, Bristol Temple Mead, having worked in on a service from Paddington, 1st November 1973. I think the Westerns were the best looking of all diesels.

It was around this time that I realised there was a logical progression to trainspotting. Firstly, you write down the numbers, but then you realise just having a number is not enough. You need a book which lists all of the train numbers you can possibly see, which becomes an essential reference book for two reasons: firstly, to cross off (or underline) those engines you have seen, and secondly, to find out how many more you need before you've ticked off all the engines in the class.

The eventual goal is to 'spot' all of the engines in the book - cue Ian Allen and his famous collection of trainspotting books which quickly became indispensable, two of them in particular. The first one was the ABC combined volume (listing all engines, diesel multiple units, electric multiple units and preserved engines), and the second was the Locoshed book which listed the depots where each engine was allocated.

When you first realise that the challenge is to see every single engine, it can feel overwhelming. People who had seen them all, including the small shunting engines that tended to stay in local freight yards all around the network were hugely respected.

It should be remembered here that in the 1970's, there were over four thousand engines compared to around five hundred today. I purchased my first railway book, the 1971 Locoshed in Woolworths, Broadmead, for fifteen pence, and that was it, I was on my way!

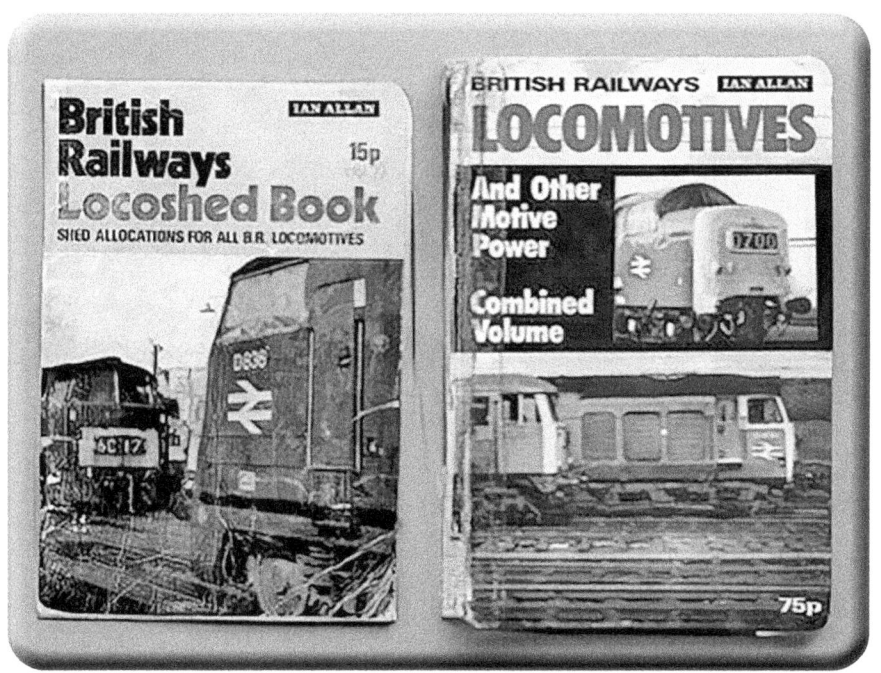

My first two railway books, the 1971 Locoshed book purchased from Woolworths in Bristol's Broadmead shopping centre and my first 'Combined Volume' purchased from the W H Smith bookstall at Bristol Temple Meads station. Both looking a bit battered.

I should perhaps say that as I became more engrossed with the hobby a whole new language opened up for me. A loco that I had not seen before was a 'cop', seeing all the engines in a class was known as 'clearing the class', going around an engine shed or depot (I use both terms throughout the book although **shed** is more associated with steam engines and depots with diesels) without permission was known as 'bunking the shed'. Going on a trip visiting many sheds was known as a 'shed bash'. In addition to these terms the engines themselves had a whole host of nicknames that I refer to but have tried to also cover in the associated glossary.

In the early 1970's, there were many classes of diesel engines working on the British Rail network. Over one thousand of these were the class 08 shunting engines which could be seen anywhere in the country, often pottering around in small yards, shunting wagons. The Brush class 47 (they were manufactured by Brush – hence the name) was the largest class of mainline engines totalling over five hundred, and again could be seen almost anywhere.

More local to Bristol were Western Region-based engines, notably the majestic looking class 52 Westerns (there were seventy-four in the class, all named Western something, e.g. Western King, Western Queen etc.) and Warships all named, not surprisingly, after warships. At this time the Westerns hauled the prestigious fast express trains between Bristol and London.

The Peak class 45 and 46 engines, of which there were nearly two hundred, (called 'Peaks' because the first ten were named after mountains in England, Scotland and Wales) primarily hauled cross-country passenger services from the South West to the North East and North West.

The lower-powered class 35 Hymeks (named after their hydraulic power transmission system) were also common in Bristol, working passenger services to Portsmouth and local freights. There were many other classes across the network as British Rail experimented with numerous different designs in the early days of diesel power.

Apart from Temple Meads, the other location which became a regular haunt in those early days was the Constable Road railway bridge in Lockleaze, a council estate in North Bristol. This bridge crosses over the four-track main line from Temple Meads to South Wales and Gloucester, just after the steep climb up Ashley Hill and provides fantastic views. There is also the bonus of being able to see the colour light signals, which cover both directions; the signals on the down line, heading towards Bristol, being south of the bridge and the signals on the up line, heading towards Filton, to the north of the bridge.

Dad had recently bought our first car, a 1966 Singer Chamois, registration JAE221D. Bizarrely I can remember these old numbers but not my current car registration! The Chamois was basically an upmarket Hillman Imp with added extras, such as a walnut veneer dashboard, and was a boxy looking car with an 875cc engine in the rear and also a pivoted rear window that you could lift, a sort of early hatchback. The purchase of this car meant that family holidays and days out became a lot easier.

The bridge at Lockleaze was the closest place to our house, where you could see a decent number of trains, and it could be reached by car in ten minutes. On one Saturday, 25th June 1971, Dad and I were parked at the bridge. Dad was sat in the car reading the Bristol Evening Post while I scanned the horizon, watching out for an approaching train which was normally indicated by the colour light signals changing to green.

We had been there for about an hour when Dad gave the ten-minute warning for us to go home. We had already seen a steady stream of class 45s and 47s, and the lights had all been on red for some time, so I was pretty happy, but just as we were getting ready to go, the down main light changed, and visible in the distance coming from Filton towards Temple Meads was what I thought was another 47.

As it came closer, though, I was not so sure. The engine was hauling a few mark one coaches and making quite a racket as it burbled past us under the bridge, so I jotted down the number, D6543, but had to wait until I returned home to work out what I had seen. It was my first class 33, a type of engine which at that time was fairly uncommon in the Bristol area. It turned out that this was a regular working for class 33s, though and a few weeks later, I saw D6548 on the same turn.

I loved going to that bridge and would pester Dad every weekend just to snatch an hour up there. I was only ten, and it was about four miles from home; in later years, I would cycle there, but at that age of ten, it was just a bit too far.
The best times at the bridge were when the London to South Wales line was blocked due to flooding in the Badminton tunnel, which resulted in all traffic being diverted through Bristol. On those days, you were guaranteed an almost constant procession of trains passing by. At peak times, it was not uncommon to have three trains passing at once, which could be frustrating as you would invariably miss the one in the middle. Today there is much less variety to see but just driving over that bridge brings back many happy memories.

As I mentioned earlier, Dad served in the Royal Navy, and he continued to maintain an interest in all things naval. As a result of this, my first 'serious' spotting trip was a visit to Portsmouth on the August Bank holiday of 1971 for one of the 'Navy Day' events when the dockyard (and a good selection of naval vessels) would be open for the public to see.

Warships, the diesel-hydraulic variety, were affected by a change in British Rail traction policy which began to favour diesel-electric locomotives. Generally, diesel-electric locomotives, although heavier than diesel hydraulics, were more powerful and reliable. This meant that the Warships were taken off main line work and were used on secondary duties, such as stone trains from the Somerset quarries. Therefore, as we passed through Westbury (which was the main yard for marshalling stone trains), we were unsurprised to see three Warships, Foxhound, Dragon and Magpie, stabled in the busy sidings.

I wish now that I had kept a record of the engines which hauled the trains we travelled on that day; it would probably have been a class 35 Hymek, which was a class that would soon be phased out like the Warships. Though I regret not keeping haulage notes now, it is worth pointing out that at that time, there did not seem to be the same level of interest in recording how many miles you could travel behind different locos. This aspect of the hobby seemed to develop more in later years.

As we pulled into Portsmouth and Southsea, D6547, (another class 33), was stabled at the station, fantastic! I could now cross off three class 33s in my ABC, two of which were consecutively numbered. Such things were important.

Portsmouth Harbour Station lives up to its name and is built on stilts literally over the harbour. The southern region four-carriage unit (4 COR was the technical term for the unit with the COR standing for corridor meaning there was a corridor running through the carriage) stabled on the buffers that day dated back to 1937, but nearly thirty-five years later (when Dad and I visited), it was still plying its trade on the south coast. It was clear to see that the sense of history at this location also extended to the railway scene.

By the time Dad and I visited in 1971, the 4 COR units had been relegated from express work, though and were only being used on slower stopping services. The 4 CORs were designed by R. Maunsell and originally built for the Portsmouth electrification; previously the Southern Region had favoured running two six car units on services forming a twelve carriage train. The new 4 CORS meant that three units could be coupled together with the corridor/gangway between the units enabling passengers to move freely through the entire train, something that was not possible with two six car units. This provided greater operational flexibility. Constructed at Eastleigh Works, the four-car units, originally painted in Brunswick green, were the mainstay on the London to Portsmouth route for many years.

As with most Bank Holidays, the weather was a let-down, gloomy and unseasonably cold. There were a lot of ships on display in the dockyard at Portsmouth, including the County class destroyer HMS London and a few Leander class frigates, including HMS Hermione.

Also visiting was HMS Endurance which ten years later would be remembered in history for her service during the early days of the Falklands War. It had been a great day out.

Inevitably, the first numbers I crossed off in my ABC, were engines which visited Bristol on a regular basis. There were, however, visits from locos based at other depots, for example, Peaks from Leeds Holbeck and Newcastle's Gateshead depots, 47s from Crewe and occasionally a class 37 from Cardiff Canton shed. These 37s were predominantly used for freight in the Western Region, particularly for coal workings in South Wales.

The first 37s I saw were D6606 and D6985 on 3rd July 1971, both of which were coupled together ('double headed') to provide greater power and working a train of pulverised fly ash from Aberthaw Power Station.

The ash was a by-product of the power station and was being used in the construction of the M5 motorway in Somerset, meaning these trains were a regular feature throughout the summer of 1971.

It was clear from my Locoshed book that many 37s were allocated to the two main depots in South Wales - Canton (Cardiff) and Landore (Swansea) - so at some point I was going to have to travel further afield from Bristol if I was going to increase the crossed-out numbers in my ABC. As I was still young, these early trips were with Dad, who seemed to get as much out of them as I did.

On 9th October 1971, we went over to Severn Tunnel Junction to see one of the most famous steam engines GWR King Class No. 6000. This engine - named King George V - was hauling one of the first steam trains to run on British Rail since the end of steam in 1968.

I'd never been to Severn Tunnel Junction before, and in my mind, I'd visualised an underground station (no doubt influenced by our visit to London earlier in the year and my first trip on the tube). The journey from Bristol to Severn Tunnel Junction was in a Swindon-built cross-country diesel multiple unit (DMU), which was resplendent in the British Rail corporate blue and grey livery (the BR colour scheme at the time) and had two big square windows on the front.

The unit swayed and rattled as it built up speed going down the incline towards the Severn Tunnel before sounding out a loud blast on the horn and plunging into the dank-smelling gloom of the tunnel. Somebody had left a window open, and the noise was deafening as we bounced around inside the carriage. After what seemed like an age, we emerged from the tunnel into broad daylight and then coasted into the station at Severn Tunnel Junction. There were a few hours to wait before the steam special was due to arrive, although there was already a good-sized crowd of enthusiasts milling around, positioning their cameras and tripods.

From the platforms, I could see the large marshalling yard to the west of the station, where several engines were stabled. We walked out of the station to try and get a better vantage point for seeing the numbers. There were some other locos in the yard as well, including my third class 25, D5182, which had recently been reallocated from the Midland Region to the Western Region to enable the withdrawal of the Hymeks (as referenced above) to take place.

We made our way back to the station, which was now filling up with enthusiasts. I had borrowed Dad's Kodak Pocket Instamatic camera and was happily clicking away, but not always with the best results, in one image a man's head occupied half of the frame alongside King George V, whose steam was visible just above the chap's wispy side parting. Looking back, I find myself wondering who was this man and where is he is now? Was he reliving his boyhood spotting days?

A souvenir ticket commemorating the first time a mainline steam engine had run on BR since the end of steam. I went to see it with Dad on 9th October 1971 at Severn Tunnel Junction.

In fact, many of my early photos have either someone's head or one of my fingers creeping into the shot or, another favourite, the small camera carrying strap, which had a habit of looping into the frame just as I was about to press the shutter.

Sometimes, caught up in the excitement of the moment, I would be a little premature and press the shutter button too early, so I also have a selection of photos which appear to show an empty section of track until, when you look a little closer, you can just about see the front of an engine creeping into the frame. That afternoon though, all the interest was in King George V, buffed up and gleaming; looking every inch the grand old engine that she was. Even the commemorative bell on the front, given to her following a visit to America, had an extra shine, but for me, the sight of class 52 diesel hydraulic D1052 Western Governor roaring noisily through the station on a Paddington bound express was more exciting.

The King eventually eased out of the station and headed off towards Cardiff, an historic engine from a bygone age. Following her departure, the crowds left, and the station returned to normality. Dad suggested we take a train to Newport before heading back to Bristol, a journey which took us past the large steel complex at Llanwern and the freight yard at East Usk before pulling into Newport High Street Station. Here we saw a few more class 37s and 47s, and, despite it being a Saturday, there was still a considerable amount of freight traffic moving about.

The journey back to Bristol was quiet, though when we checked on Bath Road depot to see if there were any new locos, D5809 and D5812 were noted. These were the first class 31s to be allocated to Bristol, which represented another nail in the coffin for the Western Region's non-standard diesel-hydraulic fleet. These two locos looked distinctive with their yellow ends and green body sides with two parallel white stripes running the length of the engine. For me, this was the best of the early diesel liveries, which was soon to be replaced by corporate blue.

Only a couple of weeks later, on Thursday, 28th October during the half-term holidays, we were off again, this time for a railway-orientated trip to London. This was a trip of firsts; first class 55 Deltic, first class 40, first London Midland main line electrics and first North British class 22. The day started well. We were waiting on Platform 3 at Temple Meads when class 25 D7656 came onto Bath Road depot, having been newly transferred to Plymouth Laira as part of the vanguard of ex-London Midland Region class 25s preparing to spend a few years in the sunny southwest. There were no Westerns at Chippenham that day when we arrived, although a Hymek was idling in the station.

In contrast to the migration of ex-Midland and Eastern Region locos to the southwest, was the build-up of withdrawn engines at Swindon Works waiting for the cutter's torch. Class 42 Warships D852, D860, and class 22s D6326, and D6352, all languishing in the works yard awaiting their fate. We saw various other locos along the way before pulling into Paddington where we saw only one Warship under the great train shed, D827 Kelly. Amongst the engines that you would expect to see were a couple of class 31s - D5535 and D5536 - which had been transferred to Old Oak Common shed a couple of years earlier from Norwich and Stratford. These were used primarily for moving empty coaching stock between the station and Old Oak carriage sidings, a few miles west of Paddington.

From Paddington, we took the tube to Kings Cross and emerged up the steps from the subway to face the magnificent Italianate architecture of Lewis Cubitt's train shed. On the buffer stops was D9012 Crepello, my first class 55 Deltic. At the time, these magnificent, powerful engines hauled expresses up the East Coast main line from London to Edinburgh. There were twenty-two in the class, all of which were named after either racehorses or regiments. The name 'Deltic' came from the design of the engine (the pistons are shaped in the design of a triangle similar to the Greek letter Delta). After a quick dash up the platform, I saw my first class 40 waiting in the small servicing shed by the tunnel entrances, D284.

Two hundred of these powerful diesel electrics were built and spent most of their time on the Eastern, Scottish and London Midland regions, they were at the forefront of the shift from steam to diesel and were rarely seen on the Western Region, hence it being a good cop for me. This was swiftly followed by my second Deltic D9006 Fife and Forfar Yeomanry, which looked resplendent in its two-tone green livery.

Our next visit was to the impressive St Pancras station with its splendid single span roof, the largest of its kind in the world when first constructed; I saw a couple of Peaks on the buffers, although only one was a new one for me. A short walk down

A shot of eleven-year-old me taken by Dad at Kings Cross. I am keeping Deltic 55020 Nimbus company on the 10th August 1974.

the Euston Road and we were at the station of the same name, which was a shiny new icon of 1960s London, controversial and bold, its designers seemingly putting two fingers up to the traditionalists.

During the construction of the new Euston Station, its famous and iconic Grecian-style arch, which had fronted the original station, was demolished, a move not appreciated by many. The original station dated back to the early 1800s. Surprisingly some of the original 1830 stone blocks were later found and recovered from the River Lea.

Since then, other remnants have resurfaced during the ground preparation associated with the construction of the Olympic Park at Stratford in the 2000s. Pressure is growing for the arch to be reconstructed as part of a future planned redevelopment of Euston, although this does not appear to have progressed to any great extent.

In the early 70s, many of the London stations were looking tired, but Euston, following its 1960s redevelopment, was the exception. It boasted a large, open-plan concourse and impressive information boards.

From here it was a short walk down the ramps to the platforms where I spotted my first class 86 electrics, fourteen of them! I do need to make a confession here, though, I have never felt quite the same way about electric locos as I do diesels. I think this is because a diesel is a single power unit with no reliance on any other power source, whilst the electric relies either on the overhead cable or an electrified third rail.

Stabled at the side of Euston in the parcels sidings was another cop, LMR London Division-based class 25 D7650. Euston still looked new that day Dad and I visited and was impressive in a 1970s kind of way. Sadly, it has not stood the test of time, though, and today it looks as out of date as its close neighbours did back in the 70s.

Quick visits, after Euston, were then made to Liverpool Street (first box head code class 37 cop - D6750) and Victoria before we headed back to Paddington where, standing on the buffers, were the Bristol and South Wales bound Metropolitan-Cammell Pullman trains waiting to depart.

The following week, the harsh realisation of the demise of the Western Region diesel hydraulic fleet kicked home. This was a time before social media and text messages, but word soon got around that many withdrawn-from-service hydraulics were stabled at Marsh Junction sidings behind Temple Meads station. And so it was, on the 6th of November 1971, that Dad and I went to pass another hour at Temple Meads whilst Mum was shopping in Broadmead.

The afternoon was gloomy, British Summertime had just ended, and the station was quiet apart from a steady stream of spotters either heading off or coming back from the sidings at Marsh Junction. Dad and I followed the crowd and, after seeing class 46 D151 nestled at the back of Bath Road shed (viewed through the shed window from the depot car park on Bath Road), we eventually found ourselves behind the station on Albert Road. From here, it was possible to see the sidings, which were on a spur just off the Temple Meads Station avoiding-line.

There were convenient gaps in the railings through which people were accessing the sidings - it was a bit of a free-for-all with no railway officials in sight. We found twenty-nine locos stabled, all looking a bit unloved and waiting for their final tow to Swindon. I don't know how long they had been there, but over the next few weeks, there was a steady flow of locos making that final journey.

Within these twenty-nine, there was a mix of Warships, Hymeks and a few class 22s. Some of the engines I had seen before when they were still in service; however, I had never managed to see any of the North British constructed class 22s in revenue-earning action, so it was a little sad to see some of them at what was not their finest moment. It was then that I realised the iconic image of a class 22 engine hauling a small rake of milk wagons on a sleepy North Devon branch line would be no more.

Six months after that first visit to Severn Tunnel Junction, on 4th April 1972, we were in South Wales again, and this time we went further west to Cardiff.

My first impression of Cardiff was that it was really busy, with an almost constant stream of DMUs (diesel multiple units) coming and going, together with a steady flow of freight. I knew there was a large diesel depot at Cardiff but was not sure where (at this point I didn't have the indispensable Ian Allen 'Locomotive Shed Directory' which provided information on all the depots on the network), however a fellow spotter pointed it out some distance to the west of the station. He said it was easy to walk to and that there was a footbridge at the end of a cul-de-sac which gave a good view of the yard.

Just before we began our walk to the depot, D1017 Western Huntsman brought a service from London into the station, which was followed not long after by the stylish South Wales Pullman. These eight-car units looked good with their yellow noses and grey-blue livery, so Dad and I took a moment to admire them.

The walk to the small cul-de-sac took about twenty minutes, and we found it led off Ninian Park Road to the depot. Though it had not been a long walk it was frustrating as we were able to hear trains rumbling along the main line but could not see them due to the buildings between the road and the railway. True to my fellow spotter's word, though, at the end of the cul-de-sac was a footbridge which led to the depot. The footbridge was long and spanned the main line before ending in a flight of steps which led to a three-lane servicing shed. This was a relic from the former steam shed, which had previously occupied the same site. The depot - Cardiff Canton - was the largest in the Western Region and had between one and two hundred locos allocated throughout most of its lifetime. It was the last part of the jigsaw in the Western Region dieselisation programme and did not fully open until September 1964.

I, like many others, would spend hours on that famous footbridge at Canton over the next few years, watching engines moving around the depot yard. The foreman's office was just inside the servicing shed on the left-hand side, and it surprised me how often a less than hopeful request to have a look around was granted. Canton was a mix of the old and the new; the servicing shed was complimented by a purpose-built cathedral-like four-lane double-ended shed with high-level inspection platforms, overhead cranes and all the plant you would expect in a major maintenance shed. We spent a good and very enjoyable hour on our first visit to Canton before heading back to the station.

In the afternoon, back at Cardiff Central, we were rewarded with a procession of class 37s growling through the station (hence their 'growler' nickname) on coal trains hauling rakes of 16 or 21-ton wagons, many emblazoned with 'House Coal Concentration' on their sides. D6973, D6974, D6975 and D6999 were among them.

These engines were covered in grime and coal dust, and though some were in the new blue livery and some in the old green, it was difficult to tell which was which through the dirt.

They made a fantastic sight and sound, particularly when they stopped for a red signal; the wagons clanked together, and their brakes squealed as the train ground to a halt. Then, when the light changed to green, the driver blasted the horn and pulled the power lever, allowing a throaty roar to fill the station accompanied by a plume of dirty diesel exhaust.

A view of the yard at Cardiff Canton taken from the famous footbridge that led from the cul de sac to the depot, with 47123, 25152, 25031 and 37183 on the 22nd April 1974.

During this visit, I really felt a sense of the region's coal mining history with the smell of coal dust lingering throughout the station long after the locos had departed. The journey back to Bristol provided a few more cops, with the last of the day being class 25 D7588, which had been newly reallocated to Bath Road from the London Midland Region's Manchester Division. The influx of engines reallocated from different regions was definitely an indicator to me that things were on the change.

One of the annual Navy Days on the August Bank Holiday in 1972 was at Plymouth (Devonport), and so Dad and I were off again. I was looking forward to the journey - especially as this was the first time since I had become interested in railways that I'd travelled south west of Bristol.

We had ventured south a few years earlier on a family summer holiday to Paignton, but that was before my interest had I remember, on that holiday trip, mum insisted on taking all our own bedding (I don't know if that was the 'done thing' back then), so we were all laden down with several awkward to carry suitcases. On that occasion, our journey had taken us along the stretch of line which skirted the English Channel through Dawlish. This delivered good value as we were treated to an unseasonably rough sea, throwing spray onto the carriages.

I digress. Back to 1972 and my first trip south as a fully-fledged trainspotter. We passed through the south Bristol suburbs on our way to Devon, and the first sight of any interest was the small railway museum at Bleadon and Uphill station, which had a collection of exhibits stabled in the station and visible from the train. Its star attraction was a former GWR tank engine, 1338, (it's called a tank engine as it carries its water in a tank on the engine as opposed to in a separate tender) which had resided there since 1964 (although would eventually move to Didcot, where it was restored).

At Taunton, we saw another GWR tank, 1421 Pontyberem, which had been built in 1900 for the reduced height Burry Port and Gwendreath Valley Railway in South Wales. This engine, though, spent most of its life working for the National Coal Board before moving to Didcot and then onto a private site in Barry.

It was a cloudy day as we headed further south, and whilst there had been hints of sunshine, the sea looked very grey as we passed through Dawlish, but no sea spray this time! A short while longer and we were on the approaches to Plymouth and Laira depot. Laira was the first of the Western Region's new major diesel depots and enjoyed a healthier budget than many of the construction projects which followed.

Built alongside the steam shed, it was an impressive building with its tall and majestic-looking two-lane heavy maintenance shed, but unfortunately for me, it was just a bit too far away to make out many numbers as we passed by on the train. Disappointingly, I only managed to note a class 03 shunter as we skirted around the site.

It was a long walk from Plymouth station to Devonport dockyard along the seemingly never-ending Union Street; however, we were able to spend an enjoyable few hours clambering over a good variety of naval vessels that were on display for the weekend. The long trek back to the station was made a little better by seeing Warship class D821 Greyhound heading into Plymouth station on a London-bound express – but only a little.

Our journey home was punctuated by noting down various engines at Newton Abbot, Exeter and Taunton. When we reached Bridgwater, shunter D2133 could be seen outside the British Cellophane Works. Initially, D2133 had been allocated to Taunton in 1960, but I had by now discovered that engines, during their lifetimes, could move all over the country, so I was able to appreciate how unusual it was for this Class 03 shunter to have spent its entire life in the West Country. It undertook a few spells at Bristol Bath Road and Swindon before then being sold in 1969 to Courtaulds for use at their cellophane works. Its final journey took it to the West Somerset Heritage Railway, where it remains to this day.

Throughout the 1970s, British Rail ran a programme of excursions known as Merrymakers. The destinations for these were fairly consistent: York, Edinburgh, Blackpool (for the illuminations), Dover, Norwich and usually a couple of 'mysteries' where the destination was withheld. I only ever went on one 'mystery' which ended up at Bognor Regis. Though it is a nice enough place, I never felt the need to try another mystery tour. The Merrymaker excursions took place throughout the year, of which two were usually to York, one at Easter and the other in the Autumn.

In 1973 the Easter trip to York was on 7th April, and Dad bought us some tickets. I was gob-smacked - this really was a dream come true, although, on reflection, I cannot decide whether I was simply easily pleased or that the bug had taken hold to such an extent that my excitement was justified.

It was an early start as class 47 D1587 brought the Mark 1 coaching stock into Temple Meads. I had been excited for weeks and even made a note of all the depots we would pass, courtesy of a shed directory published by the Dalescroft Rail Fans Society (this would later be replaced by the more expansive Ian Allen version mentioned earlier).

In the 1970s there was an extensive programme of excursions throughout the year. Note the ticket prices on these two running from Bristol Temple Meads

Unfortunately, I sometimes struggle to tell my left from my right, so on this day, as we passed Saltley depot just outside of Birmingham New Street on the Derby line, I found it odd that no one else was looking out of the left side window. A moment later, frantic shouts of 'stacks of locos' and gesticulations to the right alerted me to my error.

I was better prepared at Burton as D8189 became my first class 20 cop, stabled on the depot for the weekend with several others. At this time, the majority of the class 20s were used extensively for coal workings in the Nottinghamshire and Derbyshire coalfields, so were very rarely seen as far south as Bristol making these a great spot. There were over two hundred engines in this early diesel class and they had the nickname of 'choppers', due I believe to the sound of their engines, allegedly sounding like a helicopter, although to me they tended to have more of a whistling sound. They were lower powered engines (they usually operated coupled together as pairs to provide more power) and had a cab only at one end which made them a little unusual.

After Burton the slow journey north produced cop after cop, including some early diesel pioneers in the form of shunters D12108, D12136, D2038 and D2039, which were rusting away in the scrapyard of T W Ward of Beighton. They were visible from the line just north of Chesterfield. By early afternoon we had pulled into York, and within minutes Deltic D9010 'The Kings Own Scottish Borderer' pulled into the station on a London-bound express. That was my sixth Deltic, having seen the first five on two trips to Kings Cross the previous year and what fantastic machines they were!

This shot of Deltic D9006 Fife and Forfar Yeomanry arriving at York on a London-bound service was taken from the footbridge at the station. We had travelled to York on the Easter 'Merrymaker' from Bristol to York and Knaresborough on the 6th April 1974.

The two-stroke engines made such a distinctive sound under the magnificent station roof. When they were introduced in 1961 they were the most powerful diesels in the world with a power output of 3,300 hp and could reach 120 mph.

Along with a steady stream of other spotters, we headed out of the station past the Royal York Hotel and onto Leeman Road, where we passed under the railway bridge on our way to York shed which was accessed by a door at the rear of the building. It was a bit like entering Aladdin's cave as you were immediately confronted by class 47s, 40s, and 31s, all awaiting maintenance. There was never much sign of fitters working over the weekends in the seventies - perhaps there weren't the same demands on motive power as there are today and weekend working was not required to any great extent.

At this time, the old roundhouse at York, a remnant from the steam age, was still used for stabling locomotives, although it was to be shortly decommissioned and converted into the National Railway Museum. That day we were able to walk around the whole site. The semi-derelict roundhouse was as impressive as York station itself and, in the fading light of a sunny Easter afternoon, had an almost Gothic quality, with shafts of light pouring through the glazed sections of the roof. I mused this was much like a cathedral dedicated to a lifetime of steam. The rest of the afternoon was spent at York station, noting down the steady stream of Deltics and other motive power passing through.

The day after that trip (and any others) was almost as exciting as the adventure itself. I would clear a space on the table in our back room at home and spread out my notebook along with my ABC combined volume, loco shed book and a hardbound A4 book. I would then write, ledger style in this A4 book, details of the number, location and date of each new engine I had copped. Sometimes this could take hours, especially if the trip involved lots of DMUs (diesel multiple units).

1973 was not a great year for the United Kingdom and was dominated by tragedies. IRA bombs exploded in London and Manchester, industrial accidents claimed the lives of miners at Lofthouse and Markham collieries, and the country was facing a new economic recession. It was against this backdrop that, in the first week of the school summer holidays, Dad took me up to Crewe to get a taste of action on the West Coast main line.

First view of York shed as you entered through the side door on Leeman Road, with 37114 in the far lane. 12th April 1975.

We waited at Bristol Parkway and were rewarded with class 47 D1931, rumbling into the station to take us as far as Birmingham, where a class 86 electric, E3169, took over for the rest of our journey to Crewe. It was great to spend a few hours at the busy station at a time when pairs of class 50s could still be seen hauling the Glasgow to London expresses as far as Crewe. Once there, they would be exchanged with electric traction for the rest of the journey south. On that particular day, D429 and D420 arrived in tandem and made a fine sight as they uncoupled and swapped with a class 86. It was a hive of activity, and I loved it.

I kept an A4 book of all my sightings, class by class, completed with an ink fountain pen.

The large diesel depot to the south of the station was also incredibly busy, with class 40s, 24s, 25s and 47s constantly moving on and off the shed and the adjacent holding sidings. Times were changing too, and whilst many locos had their pre-TOPS (Total Operations Processing System) number, the new order was also arriving. Electrics 85018, 81005, 81001, 86001 and 86201 were all noted displaying their new numbers and as trainspotters, we began to realise how much of a nightmare the whole loco renumbering exercise was going to be. Most classes were quite straightforward and followed a logical sequence, e.g., D6701 became 37001 (37 being the class number), D9001 became 55001 etc.; however, some classes, like the 45s, were completely random. Then a few years later, further renumbering occurred as engines were modified and sub-classes were created. Some class 47s were to have five different numbers in their lifetime!

It was a gloriously sunny and warm day and after a few hours we headed down to Birmingham New Street before eventually, class 45 D49 'The Manchester Regiment' hauled us back to Bristol passing class 37s D6931 and D6954 on Lickey banking duties at Bromsgrove, on the way. It was only a few months after this trip that the class 50s started cascading to the Western Region and became an everyday sight in Bristol.

Bristol Bath Road Depot

Bristol Bath Road was my hometown shed. During my school days, I would spend hours on the end of Platform 12 at Temple Meads watching Westerns, Hymeks and Warships moving on and off the shed. Class 25s, 50s and 31s soon replaced these - in fact, over the years, it was possible to see examples of most diesel types passing through this shed.

The introduction of High-Speed Trains (HSTs) marked the beginning of the end for Bath Road, though, and through the early 1990s the number of locos using the shed began to decline. In 1995, the only locos which regularly visited were the red-liveried Rail Express Systems class 47s until they eventually moved off to the new servicing facility at Barton Hill, just to the north of Temple Meads.

Bath Road officially closed later that year, which, when I first started trainspotting, was inconceivable. The sidings that had once housed Westerns and previously Kings and Castles were lifted, and the site was prepared for a new lease of non-railway life.

In 1996, I achieved what my younger self would have considered a major achievement. I secured a new position at Bristol City Council, which just so happened to be based in an office that was almost within touching distance of the entry/exit lines into Bath Road depot. I could look out of the window by my desk and have a clear view of the station's east side.

Sadly, this window seat came one year too late (given that the depot closed in 1995), although I guess I wouldn't have got much work done if the depot had still been open for business!

I could, however, see the lines heading south through the far side of the station which resulted in one or two freights passing through each day.

Pairs of class 37s hauling the distinctive silver wagons that were used for china clay traffic (known as silver bullets) were the highlight of that time. To my utmost dismay, six months after starting the job, a series of hoardings were erected along the length of the unused platform on the eastern side of the station, which completely blocked off my view. These did get taken down sometime later, but by then, I was working in a different office.

The first engine shed at Bath Road, the Bristol and Exeter Railway workshops, opened in 1852, with the site subsequently being redeveloped in 1934. The earlier buildings were replaced by a ten-road main shed and a three-road heavy maintenance repair shop.

In 1960, the shed closed to steam engines and, following further conversion reopened in 1962 as one of the main Western Region diesel depots. The former heavy maintenance shed remained, but the main shed was converted into a six-lane diesel maintenance facility and a new three-road servicing shed was constructed.

A few years ago, I came across a copy of the plans for the construction of the diesel depot, and it was interesting to see that the original drawings were slightly different to what was finally built. The initial proposals included a locomotive standing shelter facility, like the one constructed at Landore (Swansea) which was to have been in front of the servicing shed. The administration block was also planned to stand a few stories higher than it actually did.

The Ian Allan British Locomotive Shed Directory describes Bath Road depot as being *'on the east side of the line at the south end of Temple Meads station; the yard is visible from the line.'* I love those entries in the Directory; the format is the same for every shed: a simple geographical positioning statement followed by a set of instructions of how to get to the shed by either walking or using public transport. Perhaps reflecting the different times, there were no driving instructions.

Phrases such as *railway over-bridge* and *railway under-bridge* were frequently used in the descriptions – presumably because simply saying *railway bridge* was not precise enough. Cinder paths, as opposed to footpaths, were often mentioned, which added a sense of romance, and flights of steps leading from bridges (Southall) or footbridges were found at the end of cul-de-sacs (Cardiff Canton). There was even a tunnel that had to be negotiated (Stratford).

Browsing through a second-hand bookshop in Bristol recently, I came across the forerunner to the Ian Allen Directory. Compiled and published by R.S.Grimsley in 1947, it was in the same format and used the same descriptions for each shed as in the later Ian Allen version, which must have replaced it. I found this to be a fascinating read, so I have set out below a paragraph from the preface which captures the time perfectly.

> *During the war years, the greatest drawback to the enthusiast was the total restriction placed on visits to the various railway running sheds and workshops. Since the cessation of hostilities, the railway companies have, as far as circumstances permit, reinstated such facilities, and it is only to be expected that all students of the locomotive wish to take advantage thereof. The newcomer to the hobby, the serviceman with years of overseas service, the war worker too busy for much wartime observation, all wish to make a more practical connection with their principal love, the steam locomotive.*

The guide ends with the obligatory warning that it does not provide any authority for you to enter the premises.

There were four ways into Bath Road depot. The official entrance was from the layby off Bath Road after the railway bridge – sorry, railway under-bridge. Another was on the opposite side of the road to the main entrance, and this was mainly used by vehicles as it led under Bath Road to the back of the main shed by the turntable area. The turntable was used not only for turning locos but also for lorries and other vehicles due to the tight confines of the site.

A third point of access was via a flight of steps on Albert Road that led onto a bridge carrying the Temple Meads avoiding line (behind the back of the depot), and the final approach was 'route one' across the boarded crossing at the end of Platform 12 at Temple Meads.

The first Locoshed Directory produced by R.S. Grimley in 1947, an early version of the later Ian Allen publication.

If you walked up Bath Road to the depot entrance and entered the small car park, you could stand on its low perimeter wall and see through the railings into the first lane of the maintenance shed. Here, there were often a couple of locos to be found. By turning right through the entrance door (which was three floors above ground level), going down a corridor and then descending a flight of stairs, the windows in the stairwell would give you a view across the front of the maintenance shed.

If you continued to the bottom of the stairs, you would find yourself next to the first lane of the shed. At this point, you could bunk (enter without permission) around the shed (if you felt brave) or return up the stairs and leave the site. The other option was to proceed through the swing doors into the large driver booking on/off hall, approach the foreman's window, ask to have a look around and usually be told to *!&% off. Then you would need to leave!

For the record, I have never bunked around Bath Road, probably because it didn't seem worth the risk to see locos that I had more than likely seen before.

Bath Road shed had a reputation for being a difficult depot to get around and had adopted a policy of not issuing permits to railway enthusiasts. This appeared to be a unilateral decision taken by Bristol as permits were regularly issued for other Western Region depots. Ironically, for a shed that seemed to adopt a zero tolerance to spotters, the main entrance for many years hosted a selection of black and white postcards in a display case depicting examples of the different classes of locos allocated to the shed over the years. These postcards could be purchased, though how you lawfully gained access to buy these, I never managed to discover.

The final Open Day at the depot was in 1991, and was also the first for many years. The previous one had been held in 1970, just before I picked up the railway bug. In the mid-1970s, Modern Railways magazine announced the date of a proposed Open Day at Bath Road, which turned out to be untrue. However, this did not stop a hundred or so enthusiasts descending on Bristol on the proposed date, resulting in a cat-and-mouse game with the shed staff. Spotters would keep running off the end of Platform 12 at Temple Meads into the depot complex, hotly pursued by the foreman and any other staff he could muster. This pantomime continued all afternoon until the appearance of the Transport Police finally calmed things down.

The first time I went around the shed was on the 21st of June 1973 and came about through an unexpected source. At school, I had become friends with a boy whose father was a driver at Bath Road. It didn't take long for me to recognise the potential opportunity that came with this friendship, and eventually, I asked for the obvious favour. A few weeks later, on a grey and dismal Thursday afternoon, I waited with my friend at the entrance to Bath Road. After what seemed like an age, his dad turned up and gave us a tour of the shed. He took us into the cab of D1046 Western Marquis, which was stabled outside the heavy maintenance shed, and even let me sound the horn.

There were twenty-seven locos on the depot, which was about average for a weekday. There were also two withdrawn 1957 constructed Park Royal DMUs (diesel multiple units) in the yard, which had been at the depot for several months. (In case anyone is thinking why I wasn't at school, we were given Thursday afternoons off school to compensate for having to go in on Saturday mornings. In hindsight, this was a poor trade-off for all those disrupted weekends). After that first visit and over the years, I would say I probably only ever went around Bath Road shed four or five times.

Back to the (final) Bath Road Open Day in 1991 which was a joint event with the nearby HST (High Speed Train) depot at St Phillips Marsh and marked a few years since I had visited any depot.

My old spotting friend Brian and I met up for this event – he had come down from London – and we enjoyed a few pints in the Bell on Redcliffe Mead Lane, by Temple Meads station, before going to the Open Day. It was great having the time to catch up on all the news as well as reminisce about our old railway trips, back in the day.

It had been a while since we'd attended an open day together, so our expectations were high. There was the usual collection of ancillary events at the open day, too, although this was the first one I have ever attended which had a display of vintage motorcycles.

A programme from an earlier Bristol Bath Road Open Day on the 23rd of October, 1965.

Brian and I were pleased to see a good variety of visiting locos; the majority were preserved engines that had been lovingly restored to working order by preservation societies following the end of their life on British Rail.

A key theme of this Open Day day was a display of locos which had been associated with Bath Road depot through the years; hence, there were Westerns, Hymeks, a Peak, and some steam engines, including Nunney Castle and the City of Truro.

The last Open Day at Bristol Bath Road took place on the 29th June 1991. Here is a general view looking towards the six-lane maintenance shed with 47834, D120, D1013 and D7018 visible.

The Bath Road Open Day on the 29th of June 1991 was also shared with St. Phillips Marsh Depot, where 50015 is seen on display.

D1013 Western Ranger at Bath Road Open Day on the 29th June 1991.

A Deltic and a Class 91 electric, both thoroughbreds associated with the East Coast main line were also there. They did not really fit in but were still appreciated.

Ironically, just a few months before its closure, the depot was awarded ISO 9000 accreditation (for Quality Management), and the event was recorded in a video by *Mainline Productions, 'A Day in the Life of Bath Road'*. This video not only takes the viewer on a visual tour of the depot but also covers various loco movements throughout a night shift.

The ISO 9000 award ceremony had more than a twinge of 'the end of an era', and 47816 was named after the depot in recognition of Bath Road gaining the award.

By the early 2000s, though the depot buildings remained intact, the track had been lifted, and the area in front of the main shed was now being used as a building materials store - in conjunction with works being carried out at Temple Meads. I made a few visits to the derelict site during this period and entered the complex via the vehicle access route on the opposite side of the main entrance on Bath Road. I was surprised to find this unsecured.

47709, 47568, 47467 and 47780 stabled outside the maintenance shed at Bristol Bath Road on the 9th of June 1995, just three months before its closure and the end of an era. Gone but not forgotten!

It seemed strange walking around the site. The maintenance shed and heavy repair shop were locked up, but there was an open door giving access to the three-lane servicing shed. It was melancholic standing in the deserted, silent building, thinking of all the engines which had passed through. There was a hint of the Marie Celeste about the place as the offices appeared to have simply been abandoned; tables and chairs were scattered around along with the usual detritus left behind when a workplace has stopped functioning.

Around this time Railtrack obtained approval to dispose of the site, despite an objection lodged by a Bristol MP who raised concerns that redevelopment was premature due to increased rail usage. A few months later, though, demolition contractors moved onto the site.

During one of my (clandestine) visits, I asked if I could have a look around the locked-up main sheds. Following a few phone calls by the on site contractor it was arranged and a week later I was given accompanied access to the site. I took video footage and photos of the depot buildings and even managed to rescue several chairs from the offices to replace the poorer quality ones in our small mess area at work. The two main sheds still smelled of diesel fuel and oil, although the only visitors to the site now were pigeons. The surrounding sidings had all but given up the battle against waist high weeds. As we walked around, I explained to the young contractor what it was like in its heyday, pointing out the under-frame cleaning pit (which was gradually disappearing under vegetation), the jacks in the lifting shop and the static engine load bank formally used for testing engines. He seemed genuinely surprised and a bit bemused, asking me how I knew these things.

The servicing shed at Bristol Bath Road overrun with weeds and waiting for demolition. Taken in the early 2000s.

A view across the front of the six-lane maintenance shed at Bath Road Depot just before demolition. The tall glass window structure in the centre background was the stairwell that gave you a view across the front of the shed. Note the Intercity livery on the shed doors.

At first, I thought he had been impressed with my knowledge; however, at the end of the visit, he gave me a tape of spiritual Islamic songs. This was when the penny dropped and I realised he had formed a very different opinion! He suggested that I might like to listen to these songs, think about their meaning and consider where my life was heading. I took the tape, thanked him and walked away feeling a bit uncomfortable. Actually, if I'm honest, I felt a bit of a plonker.

My proudest moment, though, was getting hold of one of the hexagonal, red-painted numbered signs which were fitted above the entrance to each shed lane. I have the one that was above lane six (or is it nine?). Due to the location of my office, I had been able to keep an eye on the demolition of the buildings, and asked one of my workmates, Ian (an inquisitive ex-police inspector and railway enthusiast) if he fancied a challenge. I asked if he thought he'd be able to persuade the contractors to let me have one of these signs, and a few hours later, he proudly returned with said sign, all for the cost of a bottle of whiskey! There had been significant interest in the signs apparently, and 'lane number 1' had allegedly gone to one of the previous foremen. The whole site has now been cleared and is earmarked for redevelopment.

I remember feeling a sense of sadness when the sheds were demolished; it was as if a part of me had gone as well. It was the same feeling I'd had a few years earlier when Eastville Stadium, home of my beloved Rovers, was demolished to make way for an IKEA superstore. The place where I once stood on the terrace watching our flat-back-four trying to hold the offside trap had now been turned into a flat-pack-floor. Such is life.

As one chapter of Bristol's railway history closed, so another began, with the opening of a new loco servicing facility at Barton Hill only a stone's throw away from the old Barrow Road London, Midland and Scottish steam shed. Barton Hill has a good railway pedigree; the engine shed was converted from buildings which had originally formed part of the Bristol and Gloucester engine shed constructed in 1844. The site is quite cramped, though, and locos are serviced in a three-lane dead-end building.

My prized possession! One of the metal plaques attached above Lane 9 (or was it Lane 6?) rescued from Bristol Bath Road Depot prior to its demolition.

The depot is accessed off Days Road through a drive which leads to a car park. From here, you can view the shed. I cycled down Days Road on the way to work most days, and on one occasion, I went into the car park only to be told I had no right to be there. I was to leave immediately, otherwise the Transport Police would be called.

The following morning, I decided to try again, and instead of cycling, I took the car. I drove into the car park, wearing a suit and passed by the same official. This time, he nodded at me before going into his office. He clearly didn't recognise me from the previous day and thought I had a legitimate right to be in the car park.

Perhaps I should have worn a high-visibility vest and walked around the site. I doubt if anyone would have stopped me.

Nostalgia is good, though it can distract you from

 A less common view of Bath Road shed from above the maintenance shed, whilst it was awaiting demolition.

the present-day life which keeps us all busy. I look back fondly to the days when Westerns, Hymeks, Warships, two-tone green class 47s and others could be seen lined up in the yard at Bath Road. I, for one, miss those days and the memories of busy stations and sheds crammed with locos.

67021 and 67002 on Bristol Barton Hill depot with 66180/101/241 in the background on the 18th of March 2006. At this time Barton Hill was a busy depot mainly responsible for servicing class 67s that operated mail trains. 67002 was back in service following the accident at Lawrence Hill in 2000.

I have come to realise that the railway scene does not stand still for long. During the mid-1980s, for example, Southern Region based class 33s dominated the Cardiff-Bristol-Portsmouth services; however, more modern DMUs (diesel multiple units) were under construction to replace them and therefore, as the use of these locos would be short-lived, I attempted to photograph as many as I could. At the time, the 33s were so common in Bristol that enthusiasts paid scant attention to them, yet, within twelve months, they had been replaced by Leyland sprinter DMUs. Often, we are left with only photographs - of course, we have our memories and a list of numbers in a notebook - but oh, how I wish I'd had a digital camera back then to do justice to the scenes presented to me.

When Barton Hill opened its doors, red liveried Rail Express Systems class 47s dominated the scene with a couple of class 08s and the occasional class 56, 58 or 60. After this, class 67s became the norm when they took over the mail services, until this traffic left the rails and began clogging up the motorways.

Barton Hill shed is much quieter now, with just the odd visit from a loco dropping off or collecting stock for repair at the facility. For views of the depot, the elevated St. Phillips Causeway, a busy dual carriageway which passes over the northern part of the Barton Hill site, provides a great panorama across the approaches to Temple Meads. It's not a bad place to be stuck in traffic, although there have been several times when I have nearly driven into the back of the car in front when distracted!

Hither Green based 33048 on the down through road at Temple Meads on the 20th of December 1986.

Branching Out

In 1974, our annual family summer holiday was going to be a week at 'The Mumbles' on the western edge of Swansea Bay which nestles at the start of the Gower Peninsula. We were staying in a three-bedroom bungalow which had, according to the brochure, delightful sea views. Said brochure enticed using a selection of photos showing glorious sun-drenched beaches, exciting names such as Port Eynon and Three Cliffs Bay. Why? I wondered, would anyone want to travel to foreign climes?

The answer to that question soon became obvious.

Departure day arrived with heavy monsoon-style rain, not the type that will pass in an hour or so, but the sort that was set in for at least the next 24 hours. Dad valiantly darted in and out of the house, packing the car up while the rest of us sat in the front room looking like prisoners on death row. In addition, there was going to be a bit of a twist to this holiday as my sister Ann was now going steady with her boyfriend Don, and they were both coming with us, though they were making their own way there. Don had a rather fetching two-tone green Czechoslovakian CZ 250cc two-stroke twin motorbike which was to be their transport of choice. (I had to give it to Don. Whilst the rest of the motorcycling world had succumbed to the onslaught from the Japanese motorcycle industry, Don, who had recently parted with his Honda to acquire the CZ, was making a defiant stand by supporting the fledgling East European motorcycle industry.)

The final member of our holiday entourage that year was my nan, who was in her seventies but still very much feisty and sprightly!

An hour or so later than planned, I was wedged in the back of the car, surrounded by all the usual holiday accoutrements. Though uncomfortable, I was at least grateful for not being on the back of a motorbike, given the (ever persistent) torrential rain.

Conversation on the journey turned to the sleeping arrangements, which had seemingly caused some concern. Mum told me that Ann was to share a room with my nan (it was unthinkable for my unmarried sister to share a room with her beloved in those days), Mum and Dad would be in another of the three bedrooms, which meant I was left sharing the final room with Don. Which was okay, but I would have preferred my own space.

As we made our way across the Severn Bridge and along the M4 into rain lashed Wales, I started to think through the railway potential of this break. It would have been great to stop at Ebbw Junction, Canton and Margam on the way, but as a passenger squashed between the luggage, I had little to no say, so all I could do was watch the signs for Newport, Cardiff and Port Talbot as they flashed past the misted-up car window.

Eventually, we reached our holiday home (which bore only a passing resemblance to the promised photos) and quickly settled in before welcoming the two drowned rats that were Ann and Don. In addition to the atrocious weather, they had suffered mechanical problems with the bike and had been out in the elements for longer than initially intended!

Through my 'railway potential' musings, I had already decided that the big attraction of this holiday was going to be Swansea's Landore shed. As luck would have it, I managed to persuade Dad to take a trip up to the depot almost as soon as we had finished unpacking which meant that the whole weather/room share situation quickly paled into insignificance.

The first shed at Landore opened in 1872 and became a major player in the GWR's Neath Division before finally closing to steam in June 1961. The site was then cleared prior to the construction of the diesel depot which was opened by the Deputy Mayor of Swansea on the 3rd May 1963.

The diesel depot was formed in a triangle of lines and was situated about a mile north of Swansea High Street station. It consisted of a three road open-ended servicing shed, which was of a similar design to the servicing shed at Bristol Bath Road along with a four-lane maintenance shed. This being much like the main shed at Canton, but half the size.

The best thing (I was to discover) about Landore is that you could walk around the whole perimeter of the depot via several convenient footpaths, bridges and roads.

At the time of this, my first visit, the main draw were the cut-down cab class 03 shunters (used on the height-restricted Burry Port and Gwendarth Valley line), along with the class 37s dedicated for working on the Central Wales line (complete with specially fitted headlights) and the locally based class 08 shunters. During the holiday, I made three visits to the shed and one visit to Swansea High Street station though disappointingly, I only managed to see one 03, D2122, which I had already clocked at Bath Road. It had now been withdrawn from service and was broken up the following year.

Four years later we returned to the same bungalow on the Mumbles though we were a smaller party. Ann was now married to Don and, my nan had sadly passed away.

I was also older and more independent which meant I had the freedom to do my own thing. I managed to visit and go around Landore on several occasions that holiday - in fact, I can't remember ever being refused permission there - and because it was only me, Mum and Dad, I was able to persuade Dad to make detours to Margam, Swansea East Dock and Landore on the way there and back.

47603 County of Somerset stabled on Swansea Landore Depot on the 3rd of May 1987. It was later scrapped at Toton in 2004.

Today, Landore is a much quieter place, not least due to the effective end of coal mining and a reduction in other heavy industries in South Wales. The loco fleet has also been significantly rationalised. The new GWR class 800 bimodal units which are used on the Swansea to Paddington express trains don't use Landore (it was not designed to handle multiple units) and instead are serviced and stabled at the upgraded carriage sidings at Maliphant, just outside of Swansea station.

I made a brief return visit to Landore in July 2005 and sadly didn't see one loco. The vegetation on the surrounding pathways had grown to such an extent that I had difficulty even finding my way. I am not sure whether it was me getting older and seeing things differently but the whole area around the depot seemed run down. The one bright light on the horizon was Swansea's new football ground just a short distance from the shed. I had liked Swansea's old ground (The Vetch), but it was always felt a bit 'dodgy' visiting there as a Rovers fan.

Summer Saturdays At Temple Meads

In the 1970s, summer Saturdays at Temple Meads were special. On a good day, you could see nearly a hundred locos from the platforms, either passing through or moving around on the depot. The station was manic, it was hard to believe so many people relied on the railway for their summer breaks. Logistically, though, it was a nightmare for British Rail. They had to use freight locomotives which would usually be resting for the weekend to haul the extra passenger trains. For spotters, however, it was great! Engines that were infrequent visitors to Bristol were pressed into action, with Knottingley and Immingham based class 47s being the highlights.

For me, a typical Saturday would start with a walk down to Sea Mills station on the Severn Beach branch line to catch a service into Temple Meads. Sea Mills was my closest station and was picturesquely located on the banks of the River Avon. The double track which used to serve the station had been singled in the early 1970s which made the platform nearest to the river redundant. For much of the 1970s, the regular service to Temple Meads was provided by single-car diesel units (nicknamed bubble cars), with W55032, W55033 and W55034 being the most frequent.

On one particular bright and sunny summer Saturday, (the 2nd August 1975), I joined a small crowd of people waiting on the platform at Sea Mills. We were looking into the distance waiting to see a unit enter the section of track which hugged the River Avon, and wound around the aptly named Horseshoe Bend, about a mile out from the station.

Eventually, it came into view and made its way into the station, rattling over the iron bridge which crossed the entrance into the silted-up Sea Mills harbour. The harbour had, many years ago, served a nearby Roman settlement which had the grand name of Portus Abonae. The unit stopped at the station with a squealing of brakes, then it was all systems go as doors were opened, passengers boarded and doors were once more slammed shut. And that was it. Just like that we were off.

The guard made his way through the train issuing pink paper tickets which he reeled off a machine that hung around his neck on a leather strap. After a few minutes we plunged into the 1,738-yard-long Clifton Down Tunnel before emerging once more into daylight at Clifton Down station. This was the only passing point on the line between Sea Mills and the main line at Narroways Hill Junction.

From Clifton Down we came into Redland before joining a viaduct which crossed the busy Cheltenham Road full of Saturday morning shoppers. After the viaduct, we entered the exotic-sounding Montpelier station, which has always been pronounced by Bristolians with a heavy west country 'er' at the end. Through another short tunnel, the line curved around to join the main line where there was always a good chance of being overtaken by a southbound express racing past along the down-fast line.

On our journey went stopping at Stapleton Road, Lawrence Hill and eventually into Temple Meads - usually arriving at Platform 1 which was a small bay platform close to the main station entrance. Platform 1 on a summer Saturday was a challenge, though. Ideally, you wanted to alight at Platform 12 as that was the one closest to Bath Road Shed, but unless you had a ticket for one of the services stopping at the platforms on the far side of the station, access to Platform 12 was restricted.

The barriers were manned by station staff who were fighting a never-ending battle in trying to reduce the number of people making their way to and from the busiest platforms. Sometimes, though, you could sneak through the barriers or alternatively try to blag your way across.

"Sorry I haven't got my ticket," I would say in my best innocent voice, *"but my Dad asked me to get a newspaper from the kiosk. Can I please go through?"*

This was my favourite 'blag'. Sometimes it worked, sometimes it didn't.

On these Saturdays, Bristol Temple Meads would be heaving with people trying to make their way onto the platforms or pushing through to the station exits. Announcements on the station's tannoy along with guard's whistles and people calling to each other created a cacophony of noise layered over a backdrop of a constant stream of trains rumbling into the station. It was trainspotter heaven!

W55034 on the Ecclesbourne Valley Railway on the 21st of April 2023. This unit was a regular performer on the service from my local station at Sea Mills to Bristol Temple Meads.

Usually, the timetable meant that morning activity was dominated by trains heading south, and then later in the afternoon, they would come back through again to return north.

Most cross-country services were hauled by class 45/6 Peaks and 47s, whilst the London services were now in the hands of class 50s. These had mostly (apart from a few still based at Crewe) been transferred down to the Western Region to take over the top link work from the 47s, which in turn had replaced the Westerns.

One of the more unusual summer workings was a Birmingham to Weston service which for several years was made up of a pair of three car Tyseley based DMUs - usually the distinctive looking Metropolitan-Cammell types. As mentioned, the Westerns were now in decline, with about a third of the class withdrawn, so only five were seen on this particular day and the influx of 25s was noticeable with a total of eight being seen. The distinctive liveries of the 1960s and 70s were disappearing fast and corporate blue began to dominate the scene. I was rewarded, though, with seeing 47094, 47138 and 47262 still carrying the attractive two-tone green livery with full yellow ends.

50018 at Temple Meads on a service from London Paddington on the 25th of May 1974.

The services from Bristol to Portsmouth were now in the hands of Southern Region, Eastleigh built, two carriage multiple units which had displaced the diesel hydraulic Hymeks a couple of years earlier.

Bristol Bath Road allocated 31128 on the up-through-line at Temple Meads on the 20th of July 1979. Still operational with Nemeses Rail. Note the brake tender behind the loco.

By the early evening, the frenetic activity of the day had subsided and I had noted a total of seventy locos, though only one cop - 47335 from Bescot. This underlined the need to start travelling away from Bristol if I wanted to get more cops.

Open Days are the big rail events of the year. They are the culmination of much planning and organisation and usually result in a cash benefit for a local charity. For the railway enthusiast, it provides an opportunity to have a good look around a shed or works and see guest locomotives that you would not normally expect at the location. Occasionally, there may even be a locomotive naming ceremony.

I have always found Open Days to be a little too polished, everything is too clean and tidy which, for the die-hard enthusiast, perhaps delivers a lack of authenticity. I think the fact that the shed or works is open for everyone also takes away the excitement of a personal visit where you are just nipping around on your own, particularly if you shouldn't be there!

Another development in recent years has been the stricter application of health and safety regulations at these events, which has resulted in certain areas around the Open Day sites being off-limits. A further drawback is that usually every other shed in the area where the Open Day is being held, becomes like Fort Knox. All that being said, I have been to some excellent open days and as they were commonplace in the 1970s, you had the luxury of choosing which ones to visit.

An every day scene at Temple Meads in the 1970s with Peak 45026 heading south on a cross country service, on the 20th of July 1979.

One I remember specifically was on the 20th April 1975 when Eastleigh diesel depot and the adjacent works opened their doors to the public. It was a joint affair between British Rail Southern and British Rail Engineering, with all proceeds going to the Southern Railwaymen's Home for Children and Old People in Woking, a fine institution founded in 1885 by Canon Allen Edwards. Canon Edwards had become concerned about several young girls who were orphaned when their railway worker father died, so he opened the home, which, having been comprehensively expanded, is still operating today.

At this Open Day there were several special attractions lined up for attendees, though the Open Day Co-Ordinator, Mr. Millard, in his introduction in the souvenir brochure, seemed particularly enthusiastic about the appearance of 'The Rail Queen', Miss Brenda Tomlinson. She, he reported, was due to tour the depot in the morning and then the works after lunch.

Mr. Millard expanded his narrative by stating that this was the first time in some years that a Southern Region girl had won the national 'Rail Queen' event. I think the competition and the associated language speaks volumes about 1970s Britain!

Another star attraction (although a little older than Miss Tomlinson), was a Class 9F steam loco, 92203 'The Black Prince'. The loco had been purchased by the artist David Shepherd for £3000 in 1967 - an impulsive buy, he'd said, following a successful art exhibition in New York.

There were about seventy of the class still operating at that time however, 92203 was definitely in the best condition. She would go on to work out her final BR days hauling iron ore from Liverpool Docks to Shotton steel works. On this occasion though, she worked a special train which was running for visitors to the open day who were travelling from Westbury.

Eastleigh diesel depot was constructed in the 1950s, initially to maintain the new two-car diesel-electric multiple units which were being introduced on services in the Hampshire area at that time. It was later expanded and became the largest diesel depot on the Southern Region.

The main attraction for many enthusiasts at the Open Day were the southern-based locos which stayed close to the depot. For me, the stars of this particular show were the class 07 Ruston and Hornsby short wheelbase shunters, which mainly plied their trade moving between the wharves and sheds at Southampton Docks. Class 33s and 73s were also popular, although another exhibit, D1023 'Western Fusilier', was a bit of a show stealer. And so, the Spring morning stretched into the afternoon, and it was soon time to cross Campbell Road and visit the works.

Programme from the Eastleigh Diesel Depot and Works Open Day on the 20th of April 1975.

The original works at Eastleigh were built by the London and South Western Railway on a site opposite Eastleigh Station (now the Barton Park Industrial Estate). These works were used for manufacturing the bulk of the Southern Railway's coaches and multiple units. Subsequent redevelopment and construction resulted in the works as they are today.

Many famous steam engines were built at Eastleigh Works, including the King Arthur, Schools, Merchant Navy and Battle of Britain classes, whilst in the 1970s, Eastleigh was also responsible for the heavy overhaul of all the Southern Region's locomotives. You would never see the volumes of engines here that you could at Crewe or Derby (at the time), but sometimes it was the quality that counted. On this Open Day there were about twenty locos around the site, and Dad took a rather fetching snap of a fourteen-year-old me standing in front of 73134 in the engine test area. I still have this photo today.

A general shot of Eastleigh Depot at the Open Day on the 20th of April 1975. This was one of the first Open Days I attended. D1023 Western Fusilier (seen on the far right) was the star of the show.

The day came to an end, and we headed back to the car for the run home via a slight detour to Westbury, where we again saw 'The Black Prince' having worked its way back from Eastleigh.

On the 30th August 1975, Derby Locomotive Works opened its doors to the public, and I headed up north with my good friend Brian. *The Stylistics* were No.1 in the music charts with '*I can't give you anything but my love*', but at that time, Motown didn't do it for me. The Detroit-based sounds seemed a million miles away from Bristol. For reasons which were (and still are) a mystery to me, there was a horticultural perspective to this Open Day with displays of potted plants and such like everywhere! I later discovered that flower shows were a regular occurrence at Derby open days - though I still have no idea why.

The 'headliners' for this show were a couple of steam engines that were displayed alongside a 'taste of the future'. Princess Royal class 7P 4-6-2, 6203 Princess Margaret Rose, and a 1F class 0-6-0 tank engine were on show, with the Advanced Passenger Train (APT) which had been constructed at Derby three years earlier. The APT would eventually spend most of its life rusting away in various sidings around the country before finally being rescued by the National Railway Museum.

Dad used his Kodak Instamatic to take this photo of me (I'm in the white jumper) at the Eastleigh Works and Diesel Depot Open Day on the 20th of April 1975. 73134 is in the Works Test Shop prior to release back into traffic. I have my binoculars and note book at the ready. I still have this notebook on my shelf at home!

Walking through the main workshops was fascinating, and I breathed in the unique smell associated with these places. If you're familiar with the cleaning fluid 'Gunk' (used for cleaning engines), this is a very similar smell. Today, one small sniff of Gunk and I am right back in those works and sheds of my youth.

47081 stabled in the yard at Westbury depot with a Western in the background on the 8th of March 1975.

Programme from the Derby Works Open Day on the 30th of August 1975. This was an annual event which took place for many years until the closure of the works. This day in 1975, was the first time I had visited Derby Works.

The trip up to Derby was uneventful from a trainspotting perspective, though I did cop four of six class 25s stabled at Saltley as we passed by. On arrival at Derby Midland station there was a good atmosphere, made all the better by the arrival of class 24s 24036 and 24040 on a special service for the Open Day. Across from the works in the Research Technical Centre, former baby Deltic D5901 was visible along with D832 Onslaught. The Warship would eventually be rescued from the scrap yard but there would be no such reprieve for the baby Deltic which was finally cut up two years later. There were sixty-one locos on the works, mostly Peaks, class 25s and shunters - this was pre-HST days. At that time Derby overhauled class 08 shunters (also known as gronks) from across the country, so you could always be sure of a good mix there.

The first workshops at Derby were constructed in the mid-1840s. From then the site gradually expanded as various railway companies merged and repair facilities became focused on a smaller number of sites. Few of the original work's buildings remain, although after some spirited campaigning, the North Midland roundhouse, which had been built in 1840 and latterly used as a crane repair workshop, has survived.

I went to the works Open Day the following year too and I'm still a bit grumpy that I managed to miss class 44 Peak 44003, which had been withdrawn and was in the yard somewhere. It was the only member of the class that I didn't see. To make matters worse, if you search the internet for shots of 44003 at Derby that day, several images pop up, so I have no idea how I missed it! Today, the site has been largely redeveloped and includes Derby County's football stadium, Pride Park.

Back to August 1975, and after seeing everything, Brian and I left the works and made our way back to Derby Midland station over the footbridge that linked the works with the station. I think most trainspotters that day boarded a bus out to Toton (the largest depot on the network located on the Erewash Valley main line halfway between Derby and Nottingham) to finish the day off, but being conscious of the time we decided to take a train to Nottingham Midland and check out the stabling point to the east of the station. Neither of us had been to Toton and though we were keen to do so, we didn't know if there was enough time to get there and back. We also weren't sure how much there would be to see at Toton, so on that day, we chose to buck the trend and do something different. (What we didn't know then was that there was a grass bank at Toton with a fantastic view and you could pick off many numbers, so we undoubtedly missed a potential opportunity. Remember, this was pre-internet or Google maps, so our information on that day was definitely limited).

Leaving Nottingham station then, we walked along London Road and looked down from the railway bridge for our first sight of the stabling point, which, when we spotted it, was packed with engines. Not only that, there also didn't seem to be anyone around! We made our way down the driveway which led to the sidings and carefully walked between the lines of locos. With no one around we were able to spot thirty-four engines on shed including twelve class 20s, six of which were cops. It might not have been as productive as if we'd gone to Toton, but we were pleased with our haul nonetheless.

Once we'd seen all there was to see, we took the train from Nottingham back to Derby Midland before heading home to Bristol. Helpfully, the return train slowed as we passed Burton Depot (due to signalling), which enabled us to get all of the class 20s that were parked up there for the weekend. Bonus!

Some birthdays are more memorable than others, and my fifteenth definitely fell into the 'more memorable' category. It began with a lovely sunny spring morning followed by the best present ever - the latest book from Bradford Barton entitled *'BR Diesels on Shed'*. I could not have been happier.

In the early 1970s, Bradford Barton took a brave step in publishing a range of books dedicated to diesel traction. The company, based in Truro, had already published books covering steam, and their first non-steam venture, *'Diesels on Cornwall's Main Line'* by H.L. Ford, was published in 1973 featuring shots of mostly diesel hydraulics across the county. The photo on the cover shows D815 Caradoc passing an old tin mine engine house at Scourrier on the 1V72 Manchester – Plymouth service. The book cost £2.75, and the whole range was available from George's Bookshop at the top of Park Street in Bristol, (which was the only bookshop in Bristol that had a railway section of any size). Sadly the bookshop is now a craft gin distillery and there are no bookshops in Bristol with a decent railway selection.

Back in the 1970s, though, Bradford Barton's sales flourished and soon a range of books covering diesels, often with a geographical theme (Eastern, London Midland, Scottish and Western regions), were published. My birthday gift book, *'Diesels on Shed'* was published in 1976, and was popular enough to carry a price increase from £2.75 to £3.25. One eye-catching shot was a two-page spread of Eastfield depot in Glasgow showing a selection of class 20s, 27s and 37s stabled at the north end of the depot yard.

The next day (the day after my birthday), Dad was taking me on a rail excursion to Scotland (another present), which would be the first time I had been north of the border and my first opportunity to see some Scottish-based engines. Even though it was 1976 the price of the tickets at £5.50 for an adult and £4.00 for a child, seemed a bargain.

The original Bradford Barton Diesel book plus my favourite, Diesels on Shed.

The annual programme of rail excursions from Bristol which were branded as 'Merrymakers' in the 1970s, usually included two trips to Edinburgh, one in April and the other in the Autumn; however, there was another memorable event happening on my birthday before I could even start thinking about the trip to Scotland.

Rovers were playing Bristol City at home (Eastville) in a Second Division local derby. I had managed to get a ticket a few weeks earlier, and by kick-off time, the excitement and nervous anticipation were causing butterflies in my stomach. It wasn't just the game, though; there was also the excitement and tension in the air which always accompanies a Bristol derby. City were riding high in what was then the old Second Division while Rovers were starting to slip towards the bottom of the table. City had thus become a good bet for promotion to the First Division, which no Bristol team had graced since the early 1900s, so there was a lot of interest in the game.

Whilst Bristol derbies may not have the same gravitas as those of Manchester, Liverpool or Glasgow, it was a big deal in our part of the world. Moreover, it was about holding your head high in school on Monday morning if your team won, instead of skulking around and trying to talk about anything other than football if you lost. I hoped City would stay with Rovers in Division Two as the thought of them playing the likes of Manchester United and Arsenal was too much to bear.

I took up my place amongst the Rover's fans in the packed Tote End (home end) and looked across to the opposite end of the ground to see a massive sea of red and white clothed City supporters gathered on the large Muller Road open-end terrace, nestled just under the M32. The atmosphere was electric, although, unfortunately, the game did not live up to the occasion and ended 0-0 with few chances for either team.

I had already persuaded Dad to pick me up after the game, which meant leaving a few minutes early because he didn't want to get held up in the traffic. As we drove back home I listened to the last minutes of the game on the car radio, and as the final whistle went and the players left the pitch, there was a pitch invasion which, in true 1970s style, caused mayhem before the police got everything under control.

It was an early start the following day for the trip to Edinburgh; the train was leaving Bristol Parkway at 06:10 (returning at 01:20 the following morning). Before this trip, the furthest north I had been on the West Coast main line was Crewe, but today, we would be passing depots at Wigan Springs Branch, Carlisle Kingmoor and New Yard, and probably visiting some Scottish sheds although we had no firm plans.

From a marketing point of view, rail excursions predominantly targeted two markets: the rail enthusiast and the 'Wallace Arnold' coach trip market, often characterised by older couples who may not have a car or simply did not want the hassle of making their own arrangements. They'd rather let someone else do the driving.

The two groups were generally mutually exclusive, and sometimes an air of tension was apparent. If you found yourself seated in a window seat, away from the central gangway of the carriage, you were constantly saying 'excuse me' as you eased yourself past your increasingly grumpy companions. This because you wanted to be by a window on the best side of the train when passing a specific depot or yard.

After eight or nine times of 'excusing ourselves' past other, non-trainspotting passengers - especially on a trip to Scotland, the patience of the Wallace Arnold passengers began, perhaps understandably, to wear thin. As the journey continued, you would also become increasingly aware of some of the character traits displayed by your fellow travellers.

At specific times, sandwich boxes would be opened, flasks undone and out would come salmon spread sandwiches (or similar) and tea. I remember a chaotic scene on a return excursion from York a few years earlier when the train stopped outside Saltley depot for a crew change.

Unfortunately, a coal train was blocking the view between the excursion train and the depot yard, making identifying the stabled locos tricky. This required the spotting fraternity to spend the entire fifteen minutes of the crew change walking up and down the aisle (trying to identify locos from between whatever gaps in the wagons we could find), much to the annoyance of the non-spotters.

Sometimes we spotters would need to lean over our other travellers, with (I am slightly ashamed to admit), little concern for their personal space. They were wholly disinterested in that elusive class 25 whose number was partially obscured by a coal wagon and thus, the relations between the two contingents of excursion passengers continued to sour.

During this birthday trip to Scotland, the long journey north up the West Coast main line produced a steady stream of numbers: class 24s at Crewe diesel depot, 40s at Springs Branch and Preston, together with one of the few remaining 50s still on the Midland Region, 50031, in Carlisle New Yard. When we crossed the border from England into Scotland, the sunshine - which had followed our progress throughout the Northwest, eventually gave way to windy rain and menacing clouds, with snow even being seen on the higher ground. Nothing could dampen my spirits, though, as I copped my first two 'Scottish' locos - a pair of Eastfield 20s stabled at Beattock for banking duties.

We stopped for an engine change at Carstairs with the electric loco that had taken over at Birmingham coming off and a class 47 being attached for the run into Edinburgh Waverley. As soon as the train reached Carstairs, there was a dash by fifty or so spotters straight out of the carriages to do a quick circuit of the small loco shed before re-joining the train.

Finally, as we pulled into Waverley, Dad and I joined the mass exodus of spotters leaving the train and heading for the first service to Glasgow. Dad purchased some tickets for the same service and we joined the crowd, waiting for the push-pull class 27 powered train to arrive.

The journey took us past Haymarket depot as we headed towards Scotland's second city and then, on the outskirts of Glasgow, we passed Eastfield depot before entering the tunnels on the approach to Queen Street station. At this point, word spread through the spotters that someone had obtained a permit for St. Rollox Works, the Glasgow workshop responsible for overhauling Scottish Region engines. I don't know if there was a restriction on the number of people allowed access with this permit, but I am sure any figure was comprehensively exceeded as a fleet of taxis ferried a considerable number of spotters (is there a collective noun for a grouping of spotters? A *gaggle* sounds about right to me!) including Dad and I, from Queen Street up to St Rollox.

St Rollox works was constructed in 1856 as a result of the Caledonian Railway relocating their main works from Greenock. Named after the ancient church of St Roche, which stood nearby, the works were expanded in 1882 and 1923. In 1921, as part of the Railways Act, the big four railway companies were formed out of the hundred or so companies that previously existed. This resulted in the works falling under the control of the London, Midland and Scottish Railway.

In 1968, a British Rail Workshops Division, British Rail Engineering Ltd, was set up, marking the beginning of a period of significant change on the railways. Steam power was ending, and increased financial pressures resulted in a cull of the number of workshops nationwide. In Scotland, this resulted in the closure of Cowlairs Works in 1968, Inverurie in 1970 and Barassie Works in 1972, leaving St Rollox as the only remaining workshop. From then on, it was renamed Glasgow Works, although many people still refer to it by its former name.

The buildings at St Rollox have retained their charm and sense of railway history with a particularly impressive front entrance. On the day we visited, though, this majestic sight was slightly overwhelmed by the motley collection of spotters forming an orderly queue outside. It wasn't long before someone took control, and within a few minutes, the entire party were making their way through the workshops.

The engines present that day were typical for the time and were a mix of Type 2s from classes 25, 26 and 27, with several withdrawn 24s littered around the site. One other interesting visitor was the preserved Gresley A4 Pacific 60009 Union of South Africa. This magnificent, green-liveried steam engine spent its entire working life based in Scotland and regularly hauled express services between Scotland and London. In fact, on the 24th of October 1964, it hauled the last booked steam-hauled service from Kings Cross. The A4 was withdrawn from service in June 1966 and purchased the following month by Scottish farmer John Cameron, who had a passionate interest in railways. It has remained in preservation, although it is no longer working due to a boiler defect.

After St Rollox, our next stop was Eastfield, a fifteen-minute walk away and the second largest depot on the BR network behind Toton. This shed was a prominent nine-lane structure with a heavy lifting area in the centre and had been constructed in the late 1960s to replace the steam shed on the same site. Any concerns about being refused entry at Eastfield soon evaporated when it became apparent that it was 'open house' at the depot, but we were quickly faced with another worry. An unleashed Alsatian dog was wandering around the top end of the yard. The dog appeared to be defending the remains of a derelict class 24 (24006) following its withdrawal from service the previous July. It would be another four years before this loco was eventually dismantled at nearby St Rollox. I often wonder if the dog guarded it for the entirety of those four years.

After negotiating the dog, we were rewarded with a long line of locos stabled outside the north end of the shed. These were a repeat of the view in the Bradford Barton book that I had been given the previous day for my birthday. In total, our tour of Eastfield shed provided seventy-two locos, almost all of which were cops.

As we left Eastfield, the skies darkened, and the first drops of rain fell, which added to the bleakness of this part of Glasgow, somewhat characterised by dilapidated-looking tenement blocks. We were still figuring out the best way back to Queen Street station, so we decided it would be easiest to flag down a cab to take us there. Then, it was a short hop back to Edinburgh Waverley with an uneventful onward journey to Bristol, which seemed to take forever.
Football hooliganism was at its peak in the 1970s, and disorder was seen at grounds almost every weekend. Manchester United hooligans had a fearsome reputation and whenever the team travelled away their yobs seemed to cause problems.

On the 2nd of April 1977, United were due to play away at Norwich in a First Division fixture. The 2nd April 1977 was also the day of the Bristol to Norwich rail excursion - hats off to the excursion planners for that one!

Eastfield Depot visited on the annual Easter Merrymaker excursion from Bristol to Edinburgh on the 9th of April 1977. The depot tended to be an open house on these occasions.

A Bath Road loco, 47286, was the haulage for this trip and, because I was making the excursion with Dad, it would be a mix of railways and culture. We eased out of Temple Meads station and then routed out of Bristol to Filton before joining the South Wales line at Bristol Parkway.

Swindon was always an interesting place to pass in the late 1970s, and today, there were four rusting class 24s and three Westerns waiting for the cutter's torch. Of particular interest were five ex-Southern Region class 499 Trailer Luggage Vans (682xxx range), which were in storage following the decline in their use as luggage vans on the cross-channel boats trains. These would eventually be converted for alternative use as departmental stock, with most of them used as tractor units to move stock around depots.

The rest of the journey to London lacked excitement, though Langley Oil Terminal revealed an Immingham-based 47, which brightened the trip somewhat. Our train was routed over the North London line, which gave a tantalising glimpse of Willesden depot, where a Hither Green Crompton (nickname for a class 33 – they used Crompton Parkinson traction motors) was sighted. After this, we journeyed through Stratford to pick up the Great Eastern main line.

Walking down to the buffet car, it became clear that a few United fans were on the train. One, who I recognised as a spotter from Temple Meads, had put away his notebooks for the day and sat alone in the buffet, quietly working his way through eight cans of McEwan's (lager)!

> Swindon Works had a regular open day in the 1970s. At the event on the 19th of May 1979, fresh from the works was Allerton 08534 whilst behind it was Bath Road's 03382.

The elderly Mark 1 carriages rattled along the main line, up through Colchester and Ipswich and eased into Norwich Thorpe station around midday. We joined the trail of spotters and made our way from the station to Norwich shed, a ten-minute walk.

The railway infrastructure at Norwich in the mid-1970s was extensive. In addition to the busy station and large adjoining freight yard, there was a substantial loco depot comprising a rambling collection of buildings - mostly remnants from the former steam shed. After a bit of a conversation with the foreman, we managed to get around the depot and noted twenty-three locos, including ten of the local class 03 shunters.

It soon became apparent that the number of United fans arriving in the city was increasing, and the roar from the crowd at the Carrow Road ground, sited just behind the depot, could be heard easily. Sadly, that day, Carrow Road football ground saw some of the worst rioting for many years. News footage showed United fans fighting and trying to climb over the roof of the stands to get into the ground. It was chaos.

When we walked back from the depot to the station, though we had been unaware of the scale of violence, we saw the fallout of some of the drama with several broken shop windows. We also noted over seventy United coaches parked on the wasteland adjacent to the station. There were a lot of fans around that day.

I almost forgot! There was a cultural aspect to our day which took the form of visiting Norwich Cathedral and, in particular, the grave of Edith Cavell, a British nurse executed by the German military during the First World War. Her crime had been to help hundreds of Allied soldiers escape from German-occupied Belgium. Edith's execution by firing squad generated worldwide condemnation.

On our return journey to Bristol, I looked out for the spotter/United fan I had seen earlier, but he was nowhere to be seen.

What happened to him does, to this day, remain a mystery.

London

After the family trip to London, which started my interest, it was a few years before I returned on a serious hardcore spotting trip and when I did, this time it was with Brian, (there had been several smaller-scale trips over the previous couple of years but nothing extensive.) As we got older, our trips were gradually becoming more ambitious so with heavy use of the London shed itinerary detailed in the back of the Ian Allen Locoshed directory, on the 29th of October 1977, we planned the 'big one'.

We decided to follow a circuitous route which had been developed by Ian Allen over the years, visiting shed to shed in a clockwise direction. The beauty of this aforementioned Locoshed directory was that not only did it give directions from shed to shed, but also directions from each shed back to the closest mainline station. This meant you could pick up the circuit from any of the main London stations.

The journey up to London was good and we saw the usual mix of locos at Swindon either waiting for entry into the works or to be scrapped. In the scrap yard that day was 24084, which would have to wait another seven months before it was cut up; however one of the Warship class Western Region diesel hydraulics, 818, would have to wait another *seven years* to be scrapped. It was always a bit of a mystery to me as to why it was not preserved.

The HST eased into Paddington, and we made our way across the busy concourse, dodged taxis and took the exit onto Praed Street. The first shed on our list was Marylebone, home of the Derby-built class 115 diesel multiple units (DMUs) which rattled along between the capital and the Chilterns. The shed was just to the north of Marylebone station on the right-hand side of the line and was a six-lane structure - three lanes dedicated to maintenance and the other three lanes used for stabling. The shed entrance was via a driveway off Rossmore Road, which led into the back of the shed and the foreman's office was to be found on the left-hand side about halfway up the first lane of the maintenance section.

I had visited this shed a year earlier on a gloomy November weekday evening and was surprised when the foreman had given me the go ahead to look around. Particularly since it was just before rush hour and the shed was packed with idling Derby units. Unfortunately, though, the fumes in the stabling side of the shed were so overpowering that I had to leave. My eyes were streaming and I was feeling extremely nauseous, but I did manage to get all the numbers.

On this second visit, it was a crisp October morning and the shed was very quiet. Units had been shut down for the weekend so the foreman was happy for us to go around, and we were able to note down twenty-nine numbers before heading off towards Euston where we bought tickets to Willesden Junction which was our next stop.

The concourse at Euston was busy with a mix of people all craning their necks to look at the large flickering arrival and departure board. Groups of football fans were heading for the exits and onto the next stage of their journey to whichever ground their team was playing at. Finding the platform for the service to Willesden, Brian and I settled down in one of the local class 501 electric multiple units (known by some drivers as Dartmoor units due to the bars on the windows) for the short journey up to Willesden.

The shed at Willesden was a large purpose-built structure spanning six tracks to the south of the station. Due to its design the shed was light and airy which was in stark contrast to the cramped and gloomy steam shed it had replaced on the other side of the main line to the north of the current site.

The foreman's office was one of the first buildings we came to after passing through the car park, and due to the open aspect, we were quite visible as we made our approach. Brian and I were a little concerned that we would be refused entry having made such an obvious arrival, but luckily, we didn't encounter any problems and the foreman allowed us to look around. There were only a few locos stabled outside, but inside we saw a good collection of electrics together with a few diesels, mostly Willesden based 08s and a Cricklewood based class 25, which were stabled on the far lane of the depot nearest the main line.

Leaving Willesden behind, the next stop was Old Oak Common, a ten-minute walk away down Old Oak Lane. Whereas Willesden was a purpose-built depot, Old Oak was anything but. This shed was a rambling affair which occupied a large area of land between the Paddington branch of the Grand Union Canal and the main line into Paddington itself. The original Old Oak shed had been designed by the famous Great Western engineer George Jackson Churchward and took nearly four years to build, finally opening in March 1906. It closed to steam engines in March 1965 before being redeveloped into the diesel depot. Some of the original structures had been incorporated into the new depot, (most notably the seven-lane heavy repair shop known as 'The Factory') and the Northwest-facing turntable. The roundhouse and three other turntables were sadly removed but a new three-lane servicing shed had been added to the site for the diesel era.

The approach to this depot took us down a long, sloping drive which led past the turntable area and into the admin offices where the foreman was based. This layout meant we saw a few locos before we even reached the office, which was a nice bonus. After a quick word with the foreman, he gave us the nod and we proceeded to look around, all the while congratulating ourselves that we had achieved access to three out of the three sheds we'd so far visited. I wasn't surprised not to get any cops here, though with eleven class 31s, all Old Oak-based, dominating the scene.

A shot of locos around the turntable at Old Oak Common with 31416, 47099 and 31131 visible. Taken on the 21st of June 1981.

Back at Old Oak Lane, we caught the 266 bus to Cricklewood Broadway. It was a slow journey on a red Routemaster as it crawled through a busy Willesden High Street bustling with Saturday morning shoppers. I had been around Cricklewood shed for the first time the previous year on a diversion from another family visit to Auntie Flo. On that occasion we had driven there in Dad's car which had been a bit of a hair-raising journey due to Dad being nervous about driving in London. We had somehow managed to negotiate our way off the North Circular, though, and into Cricklewood, where we eventually found the depot car park off Brent Terrace. I remember jumping out of the car and telling Mum and Dad I would be about ten minutes.

From the car park I took the footbridge which led across a track to the side of the shed, but there was little indication as to where the actual entrance into the building was located. I was beginning to worry that I wouldn't actually see any locos, for any that were present appeared to be inside, nevertheless and feeling more than a little conspicuous, I walked into the depot. It was bustling, this being a busy weekday, so I decided to ask the first official-looking person I saw, where to find the foreman's office. Unfortunately, this turned out to be located close to the door I had entered from, so the foreman ended up being the first official-looking person I found! Instantly I panicked, feeling sure I would be sent on my way without having a chance to see anything. At least if I'd had to walk further into the shed to find the foreman's office, I might have had opportunity to note a few locos. Luckily, my concern wasn't justified. The foreman was fine and simply asked me to let him know when I left the site. Excellent!

In the main shed, I found class 25s, Peaks, 47s, shunters and lines of Derby built class 127 multiple units. The shed had been surprisingly full for a weekday and turned out to be a real Aladdin's cave. It took me a fair amount of time to see everything so, when I finally returned to the car, it was to a mum and dad who were seriously fed up; their mood not helped by the smell of warm milk coming from the dairy adjacent to the depot which had begun to make them feel queasy.

Squeezing in a shed or station visit like this on a family day out was not unusual. I've always tried to fit them in during a journey (with a different purpose) which sometimes works and other times not. I did once manage to drive a carload of snoozing passengers into the back of Old Oak Common shed though and copped a class 60 stabled on one of the turntable roads without anyone in the car noticing (or the foreman). Anyway, I digress. Back to 29th October 1977 and the 'big' trip with Brian.

The bus journey from Old Oak to Cricklewood took about 45 minutes, and the weather on this October Saturday was much cooler than the previous visit, so no sickly smells from the dairy! We exited the bus on Cricklewood Broadway and walked up Brent Terrace towards the shed. Armed with confidence from our 'three out of three', we walked into the office and once again were given the okay to look around.

(As an aside, I have a Cricklewood Open Day programme dated 12 July 1969, from an event that was run jointly by the London Midland Region and 7029 Clun Castle Ltd. The introduction to this programme was written by Mr R.L.E. Lawrence, OBE, ERD, who was Chairman and General Manager of the LMR at the time who talked about modern railways and the fact that we were living in a 'changing world of rapid technical development'. Very prophetic words. By the time of writing these depot buildings have now all gone, and the 'modern' locos on display at that 1969 event, have long since been consigned to the scrap yard.)

Cricklewood depot opened in July 1960 and, along with Willesden, was one of the two main depots for the LMR's London Division. At that time, services out of St Pancras were all diesel powered, and Cricklewood was responsible for providing main line locomotives for the London to Derby, Nottingham, and Sheffield passenger services as well as freight traffic. The diesel multiple units which worked the suburban services out of St Pancras (class 127) and services to Barking from Kentish Town (class 116) were also maintained at Cricklewood.

Being purpose-built for diesel motive power rather than a converted steam shed, the ten-road depot was well-lit. Locos were serviced inside the shed on the four lanes closest to the main line with each lane able to hold up to seven engines. There were also three lanes in the middle of the building for heavy maintenance with the three lanes furthest from the main line reserved solely for diesel multiple units (DMUs).

Today, there were twenty-nine engines on shed, including two class 25s (new for me) and about forty DMU cars. After completing our visit, Brian and I left the shed and returned to Cricklewood Broadway to catch another bus, this time to Finsbury Park station. Here, we jumped off and walked down to Finsbury Park depot which allowed us to pass the small park on the left-hand side presenting us with the opportunity to glimpse anything that may have been outside.

Finsbury Park was a shed I'd visited several times and the foreman's comment was always the same: *"You can go around the shed, but don't go outside into the yard."*

This was not unusual. Some sheds had 'traditions', for example Allerton, just outside of Liverpool, would let me go around as long as I donated to a local charity. (Bonus - I also received a lapel badge from the charity). Imagine how much money could have been raised for good causes if this practice had been adopted at more sheds – especially Toton.

Opening in April 1960, Finsbury Park was just a couple of months ahead of Cricklewood and was the main diesel depot providing motive power for services out of Kings Cross. It replaced the large old Kings Cross 'Top Shed', which had maintained steam engines for over a hundred years before finally closing down in 1963. The contrast between the new Finsbury Park depot and the 'Top Shed' could not have been greater. Finsbury Park was a concrete and glass six-lane structure, which had raised platforms for working on the diesels and was much cleaner overall. Vastly different to its predecessor!

When the new depot was first planned, dieselisation had been seen as an interim measure for the East Coast main line; electrification was scheduled for the 1960s. In reality, the electrification programme did not start until the late 1970s which led to some of its originally planned features becoming problematic. There was never any facility for heavy repairs at Finsbury Park depot, for example, meaning engines needed to go to Stratford when major repairs were required. Though each lane at Finsbury Park could hold three engines, this was not ideal when locos became blocked in. Later depot designs adopted lanes which generally held only two locos to prevent this issue.

On our visit in October 1977, we found the shed housing twenty-six engines, including ten local class 31s, mainly used for moving empty coaching stock between Kings Cross and the carriage sidings at Ferme Park and other local traffic. Two mighty class 55 Deltics were also on shed, 55007 Pinza and 55020 Nimbus, named after famous racehorses, as were all the Deltics initially assigned to Finsbury Park.

Finsbury Park always seemed calm (probably because I only ever visited on a weekend) and was definitely a shed I enjoyed going to. Sometimes, you could hear the Arsenal football crowd roar from the nearby Highbury ground. As was always going to happen, Finsbury Park depot closed in October 1983 following the shift to High Speed Trains (HSTs) which were maintained at a new depot at Bounds Green a few miles away. This meant Finsbury Park depot no longer had a role and if you visit now, you'll see that the site has been redeveloped into flats.

By the time we'd finished at Finsbury Park it was late afternoon, so we headed back into central London, walking down to Kings Cross. It was a fair old hike and further than we thought, but we were able to pick up the York Way Goods depot shunters en-route and noted the locos at the back of the refuelling shed at Kings Cross - the ones we couldn't see from the platforms.

Kings Cross in 1977 was not one of the safest stations for young spotters. The previous year, I had gone to the station alone and spent some time stood on the end of the platform closest to the stabling point, trying to identify a class 40 lurking at the back of the yard. I was a young-looking sixteen-year-old, and after a few minutes, I noticed that there was an older, scruffy man nearby who kept staring at me. He came over and at first, I thought he was going to tell me the number of the 40, but I was mistaken. Instead, he offered me a fiver (about forty pounds today) to go into the toilets with him. In shock, I panicked, and the only thing I could think to say was, 'No thanks, I have a girlfriend'.

I'm not sure why I said that. Did I expect him to say, 'Oh, sorry mate, in that case, apologies for bothering you and good luck in getting that 40?'

Disconcertingly, my response had the opposite effect, and he became more insistent in his questioning. I quickly made up another excuse and dashed off to St Pancras whilst checking every now and again to see if I was being followed. Oddly enough, I don't remember being particularly scared, and I never told anybody about the incident. And I still didn't get the class 40!

As I recall, there were also two or three other occasions when similar things happened at Temple Meads, and when talking to Brian a couple of years later, he revealed that he'd had similar experiences. I guess boys hanging around railway stations were always going to be a bit vulnerable, perhaps it was a common occurrence and one that we should have reported to the Police. Times were different then, though that is in no way an excuse. We were perhaps somewhat naive and unaware of the potential true gravity of these situations.

Back to the trip. After seeing what was on the buffers at St Pancras (four 45/1s), we decided to take the Circle Line to Liverpool Street, where we popped up to the main line station for a quick look around. Resting between duties was 08531, the pride and joy of Stratford depot, painted in its earlier British Railways green livery with its white cab roof, yellow connecting rods and British railways 'lion and wheel' motif displayed on its body side. We then returned to the tube taking a Central Line service out to Stratford.

From here we walked left out of the station and then took a second left into the long tunnel, which led us under Stratford station and eventually to the shed.

The tunnel under Stratford station had somewhat of a reputation. We had heard through the grapevine (the Bristol railway crowd) that one of the foremen at Stratford depot took great delight in offering any spotters he caught entering the shed from the tunnel a choice: they could either be handed over to the police or receive a personal caning from him. I said it only a couple of paragraphs ago, but times really were very different back then.

Whilst there was no evidence to support the dubious Stratford foreman story, the thought did cross my mind as we walked through the long, poorly lit tunnel. Overhead, we could hear the occasional rumbling of trains and this, along with dripping water from the tunnel roof, increased my sense of unease.

The first thing we saw upon reaching the end of the tunnel was the **'Trespassers will be prosecuted'** signs. These were commonplace at all sheds, but for some reason, they always seemed to have more resonance at Stratford than at other depots. Even still, it was not enough to deter us, and a short walk on the other side of the tunnel led to the 1871 'New Shed', which was generally used for stabling a couple of 03s, a few 31s and, at the front of the shed, the Temple Mills yard shunters (if they had worked down from the large yard a mile or so north of the depot).

So far, so good.

Cautiously we proceeded, making it to the next stage which involved crossing the yard between the New Shed and the four-lane double-ended diesel shed (built on the site of the former Jubilee shed that was constructed in the 1890s). Once we had done this, we then had a decision to make. Should we go to the admin block, located on the far side of the shed, and ask permission from the foreman to go around? Or should we sneak around in stealth mode?

If we took the former course of action, we ran the risk of being refused permission and, potentially, being escorted off the site – this had happened to me on a couple of previous occasions and left me with little to show for my efforts. However, sneaking around ran the risk of being caught and handed over to the British Transport Police (or facing the legendary cane!) - so, did we stick or twist? After a brief discussion, we decided to go straight to the office, and luckily, all was good. We visited the whole complex, including the main diesel shed, the DMU shed and the adjoining works. Stratford was by far my favourite London shed back then (despite the trepidation of each visit!) and for me, a successful London trip relied on getting around Stratford.

In the 1980s, it became possible to apply for a single person permit for Stratford on a Saturday morning, which guaranteed a good day's spotting in the capital.

Once we'd finished at Stratford, Brian and I left the north side of the Thames and took a tube down to New Cross Gate, followed by a local Southern unit out to Hither Green. The shed at Hither Green was reached by following a cinder path, which we accessed through a gate at the end of one of the platforms.

During our visit in 1977, the complex consisted of a two-lane servicing shed and a six-lane dead-end shed, which was used for maintenance and stabling. The offices were on the far side of the main shed, which in itself was divided into two three-lane sections. Unlike some of the other depots, little had changed here since steam days, and the cinder path took us to the back of the six-lane shed. For me, Hither Green always seemed to be spotter-friendly, and I cannot remember ever being refused a visit. On that day we saw twenty locos on shed, including four withdrawn class 71 electrics, which had been in storage since the previous October. These would go on to be cut up at Doncaster two years later, having had a life span of just twenty years which was undoubtedly short relative to their counterparts. One possible explanation for this could be due to their sole reliance on electric power through either a third rail or overhead wires meaning they soon lost out to the more flexible electro-diesels, which could rely on a diesel engine where there was no electrical supply.

The heyday for these class 71 electrics was probably working the night ferry boat services between London Victoria and Dover. On arrival at Dover, the carriages were transferred to a ferry for onward travel to Calais, where a French locomotive then completed the journey to Paris Gare du Nord. What was interesting about the night ferry service was that you would regularly see French carriages travelling through the Kent countryside and there was definitely a romantic Agatha Christie feel to these trains.

All was quiet that day at Hither Green shed, with everything laid up for the weekend. The only sound breaking the silence was an occasional tick from a stabled engine.

After Hither Green, it was back on the bus for a short journey to Norwood Junction. Locos here were stabled by the carriage washing plant, which was really part of the nearby Selhurst depot complex. The bridge over nearby Tennyson Road spanned the tracks and provided a good vantage point for us to see, on this occasion, four shunters and a pair of 33s.

The actual entrance to Selhurst depot was a ten-minute walk away, and though we did not expect to get into the complex (because there was a live third rail), we thought we'd give it a go on the basis that we just wanted to see the depot's shunters which were usually stabled on one of the running shed lanes close to the entrance. On arrival, after being incredibly polite to the foreman, we were surprised when he agreed that we could go and get the shunters. By now it was dark and heading towards 7.30 pm. We would have loved to go to Stewarts Lane shed, but were running out of daylight. This was particularly frustrating because Stewarts Lane was the only London depot I'd yet to visit.

Instead, we headed back to Selhurst station and took a local service into Waterloo where there were two Class 74 electro diesels (converted from redundant Class 71s), stabled between turns. A short walk back down to the underground took us to Paddington and eventually home to Bristol.

One of the attractions of the London sheds in the 1970s was that most had particular types of engines allocated to them; Stratford had Toffee Apple Class 31s (called Toffee Apples due to the shape of the engine control lever), Old Oak Common had Westerns (and later) class 50s, Cricklewood had Toton class 45s along with its own class 25s, Finsbury Park had Deltics, and Hither Green had class 33 'Slim Jim's' (33201 –33212). These 'Slim Jim' locos were narrower than the standard 33s (hence the nickname), designed specifically to traverse the narrow tunnels on the Tunbridge Wells to Hastings line.

On our return journey that day, Brian and I reflected that this 'big outing' had been a good trip, netting us in excess of two hundred locos. Though this sounds a lot, it was a pale shadow of what would have been seen twenty years earlier, but the trips up to London to visit the sheds were always guaranteed to be good days, even if you could not get around them all. Gradually, as the number of locomotives being used to haul trains declined, so did the number of sheds looking after those engines and the railway today is very different. Many of the London sheds have now been demolished or replaced by depots designed to maintain multiple units as opposed to locomotives.

Even if it were possible to repeat a version of our 1977 trip to London (which would be unlikely due to increased security and a greater commitment to health and safety), you would probably struggle to see more than twenty locomotives, which puts our two hundred count that day into sharp perspective.

More Trips And Open Days

Seeing the New Year in when you are sixteen but look about twelve is not much fun. Some of my friends already looked like men with their facial hair and craggy looks which meant they never got asked their age when walking into a pub. Pubs and clubs were out of my reach as I had zero chance of convincing bar staff that I was over 18 and old enough to buy a pint. Though I was certain I'd be grateful for my youthful appearance later on in life, it did nothing for me on New Year's Eve in 1977 which was, to be frank, just plain boring. It consisted of Brian, me and a few cans of Sköl lager with nothing but the predictable New Year's shows on the tele. Still, everyone seemed to be having fun. Everyone that was, except us.

At a loss, we decided to go out for a stroll. The ship horns blasting from Avonmouth were mixed with sporadic bursts of Auld Lang Syne as Big Ben obviously heralded the start of 1978. The night air was cold and damp, yet despite this and our lack of evening entertainment, Brian and I acknowledged we were definitely looking forward to the year ahead.

Why? Because in the words of 80s pop combo the Lightning Seeds song, *'Tomorrow's here today'*, a new year for many spotters meant the beginning of a new annual collection. An 'annual collection' is something many spotters start (and re-start) at the beginning of each year to complement their 'all-time' collection. As a dedicated spotter, maintaining and adding to your 'all-time' collection can become increasingly challenging with few new locos of interest locally. This means having to travel further afield to add to the 'all-time' collection with any significance or regularity. By contrast, an 'annual collection' refreshes at the start of each year so provides the impetus to continue to spot locally because if you've already 'spotted' a particular loco, it doesn't matter. Everything resets. The desire to achieve a fantastic 'annual collection' can really inspire spotters, even if we aren't always able to travel further afield.

This concept is not unique to railways; bird watchers do something similar. Many twitchers even take this a stage further by having 'all-time collections', 'annual collections', and 'garden collections' which are based purely on the birds which land in their garden. Obviously it would be nigh-on impossible for a trainspotter to have a 'garden collection', but hopefully you get the idea.

At the start of my seventeenth year on the planet, my plan was to rack up a substantial set of numbers to kick off the new annual collection. To achieve this, I decided to go to Cardiff and stop off at Severn Tunnel and Newport on the way. I would be accompanied by Brian, so we met up at the end of his road early on New Year's Day 1978, both suitably refreshed after our quiet New Years Eve celebration.

To begin our quest, we walked down to Sea Mills Square and caught the 42 bus to Temple Meads (a Bristol-built green Lodekka double-decker for those interested in such things). Leaving the bus at the bottom of the incline leading up to Temple Meads station, we walked up Bath Road towards the depot, having decided to have a go at getting around. We didn't have any real expectation of obtaining permission, but there had seemed to be a bit of a softening of attitude towards spotters over the previous year.

Crossing the Bath Road bridge, we strained on our toes to see the locomotives in the shed yard. Brian and I stretched as tall as we could ... to the point where I began to wonder if the bridge had been purposely built in such a way to make it difficult to see over. Vertically challenged spotters had no chance on this bridge so it wasn't uncommon to see a grumpy, short person jotting down numbers - whilst their taller friend called them out! The small car park by the shed entrance was virtually empty - no doubt due to it being New Year's Eve - so we entered through the door and down the flight of stairs to the signing-on lobby, all the while expecting to be turfed out at any moment. We made it to the lobby where we found the foreman sat behind the little window to his office, busily scribbling on a pad. So far so good.

After a gentle cough to attract his attention, I asked if we could look around.

"What?" he replied, so I repeated my request.

Slowly, he put his pen down and raised his head to make eye contact with me, looking neither friendly nor pleased. It was impossible to determine his mood. Brian and I readied ourselves for the expected refusal.

"Of course you can," he said, *"watch your step and let me know when you've finished."*

Shocked, we gave him a broad smile and then scampered into the shed before he had chance to change his mind.

On shed there were forty locos, but no real surprises. The two resident 03s were there along with eleven 31s, including two Eastern visitors and a London Midland Region-based 25. There were also four 50s and the former Metropolitan Vickers Type 2 class 28 D5705 (TDB968006), which was used as a carriage heating unit.

D5705 was built in the late 1950s; one of twenty original engines in the class which had a more unusual early diesel design. The Metropolitan Vickers featured a co-bo wheel arrangement with a six-wheel bogie at one end and a four-wheel bogie at the other (where a 'bogie' is a chassis or framework which houses the wheels)[1]. Their claim to fame was working in pairs on the daily overnight London to Glasgow 'Condor' freight service but they were mechanically unreliable and withdrawn from service only ten years after they had been built. The Metropolitan Vickers spent the end of their working lives on local services in the Barrow in Furness area.

After saying farewell to the foreman, Brian and I made our way out of the shed and back down Bath Road to Temple Meads to take the next service to Cardiff. As the train headed out of Bristol, we were surprised to be diverted via Gloucester. Although adding extra time to the journey, it did mean we would pass Gloucester Horton Road depot.

Gloucester Horton Road was a large steam depot in its heyday; however, by the end of the 1960s, many of the buildings had been demolished. The former wheel drop shed was the only thing to remain; this being a single-lane building where steam engines could have their wheels dropped below rail level for maintenance purposes.

---

1 https://en.wikipedia.org/wiki/British_Rail_Class_28

By now, Gloucester Horton Road was only used as a small diesel servicing shed but we did see many engines stabled as our diversion took us past the shed and sidings. The sidings, though, were compact which meant we could only note down a few numbers no matter how much we craned our necks.

After Gloucester, our route followed the River Severn down through Chepstow before arriving at Severn Tunnel Junction which was our next planned stop. I'd been to Severn Tunnel many times since that first visit to see King George V back in October 1971, and the memory of her hauling the first main line steam train since the end of steam on British Rail, remained fond.

Severn Tunnel shed was a five-minute walk from the station, and conveniently, a lane accessing a rank of cottages about half a mile away ran alongside the small single lane shed. This meant that it was pretty easy to see the engines, which were usually stabled in a couple of long lines.

As it was New Year's Day (a Bank Holiday), we found everything quiet and locked down though we did manage to see twenty-three locos including the six-yard pilots plus three visitors from the Eastern Region. The remaining locos were a mix of local 37s and 47s.

Overhead, the grey clouds and drizzly rain were relentless and showed no signs of abating as we walked back to the station to take the next service to Newport. When we arrived, knowing there would be little to no bus services running, we decided to walk out to Ebbw Junction depot which was a 45-minute stroll (according to the Locoshed Directory).

The depot at Ebbw Junction consisted of a two-lane dead-end maintenance shed and an adjoining servicing shed with two through lanes. The shed opened in the mid-1960s as part of the Western Region's dieselisation programme and was built on the site of the much larger former steam shed.

Ebbw Junction was also one of the first sheds I visited a few years earlier, back in February 1974. Although it had been a weekday then, it had unexpectedly been a great day for trainspotters due to a miner's strike across South Wales. This meant that many locos were laid up on the South Wales sheds.

In February 1974 I had taken a Newport Corporation bus service to Maeglas from Dock Street close to High Street Station, though unfortunately, I left the bus a stop too early and ended up with a bit of a trek. Nevertheless it had been worth it. I had seen a good mix of locos, including five class 25s, D1200 Falcon (the sole member of prototype class 53, which had been relegated from top link services and was now working freight in South Wales - although it did seem to spend most of its time stopped up at Ebbw), and a good number of local 37s and shunters. And as if all that wasn't enough, just as I had been leaving the shed, D1063 Western Monitor thundered past on a London service. Without a doubt, it had been worth the walk!

Back to our New Year's Day antics, though, where Brian and I eventually arrived at the depot to find thirty-six locos, eighteen class 37s and eighteen class 08s. All of them were based either at Ebbw Junction or Cardiff Canton. It seemed strange and a little eerie walking between the long lines of engines when it was silent enough to hear a pin drop, especially since Ebbw Junction was a depot usually associated with freight traffic. Many of the heavy-duty steel and coal industries the depot supported were closed down for the Christmas/New Year break so the foreman had looked pleased to see us when we pitched up. We were probably the only human contact he had that shift! Nowadays, Ebbw Junction site has been cleared and is a Network rail facility supporting track maintenance operations.

Canton based class 25s D7520 and D7521 on a Port Talbot to Llanwern ore working, passing through Newport on the 22nd of April 1974.

The final stop on our South Wales pilgrimage that day was to Cardiff Canton. I had been to Canton many times and usually walked down from the footbridge into the servicing shed then asked for permission to go around. Surprisingly, this often worked, although the foreman would make a point of saying not to go into the heavy maintenance shed, which we translated into '*just be careful when you go in the big shed*'.

Brian and I entered the running shed and were given the all-clear by the foreman. There were sixty-two engines on shed and like at Ebbw, they were almost entirely local, including another eighteen class 08 shunters. We also ventured into the carriage shed and recorded over seventy DMU numbers - again finding many local units laid up for the Bank Holiday.

After leaving Canton, we returned to Cardiff Central before taking an HST back to Bristol. From there, we caught the bus back home to Sea Mills, impressed with our haul, which we both agreed had been a great start to the annual collection. If we'd had enough time (and daylight) to visit Barry, Margam, Landore, Swansea East Dock and Radyr as well that day, we would easily have seen over two hundred engines.

This trip in my tender teens took place before the decimation of the coal industry, before the recession and before all of the other economic challenges which arose towards the end of the 20th century. Today, sadly, you would be lucky to see as many as twenty engines on a weekend in South Wales.

The year rolled on, and the spotting trips picked up as the nicer weather and longer evenings arrived. The 17th of June 1978 was Doncaster Works open day and a coach had been organised to run from Bristol up to the works. Full of anticipation, I caught a late evening bus down to Temple Meads where there was then a three hour wait due to the coach to Doncaster not scheduled to leave the station forecourt until just after midnight. I was pleased to recognise a few faces at the station who I guessed were off to the open day, too, and together, we patiently waited. And waited. And waited.

Unfortunately, the longer we waited, the later the coach became and after some time we realised that all was not well. Frantic phone calls were being made by the trip organiser - presumably to the driver - but still no coach. Eventually, an hour later than scheduled, we were shepherded to a dilapidated coach, which had simply 'appeared' at the bottom of the station approach. This alternative coach was definitely not a National Express, and I was alarmed when I saw the driver who appeared to be half-asleep with a cigarette hanging out of his mouth. He also had a hacking cough which was loud enough to shake the coach.

With no other choice I boarded the run-down replacement and finally we were off, leaving Temple Meads before heading to the M5 where we stopped at the first services - I think to ensure the driver had sufficient coffee to stay awake! Later years would see the introduction of tachographs, which, given my concerns about the Doncaster driver, could not have come soon enough!

Gradually, night gave way to day, the sky became lighter, and dawn began to break as we passed through the Midlands. Leaving the M1 at the junction for Chesterfield, we bid farewell to three intrepid cyclists who retrieved their bicycles from the luggage hold before disappearing off into the North Derbyshire mist. They'd said they were going to do their own South Yorkshire shed tour and planned to link back up with our coach later in the day. Strangely, I cannot recall seeing them again, which suggests they may have found an alternative return option. Perhaps they, too, had been dubious about our 'less-than-awake driver'.

Despite the delay, we were still due to arrive at Doncaster four hours before the works was opening so the driver was persuaded to carry out a detour to Knottingley depot, which, excitingly, was a new shed for me. Forty minutes later, he carefully manoeuvred the coach down the drive and into the depot car park, where our visit rewarded us with thirteen Knottingley-based class 47s and three shunters, which were there for the weekend. Seven of the 47s were new to me, as were the three shunters, which made for a great start to the day.

Knottingley depot opened in March 1967 and was a bit unusual in that it was constructed from scratch on a new site, as opposed to being built on the site of a former steam shed. The depot consisted of a small four-lane maintenance shed with an adjoining wagon repair facility. It had been built primarily to provide locos to haul coal trains from the Yorkshire collieries to the nearby power stations. For this reason, the engines based here did not stray too far from the area, a bit like Ebbw Junction shed in Newport. It was a popular shed for trainspotters further afield but held little interest for local enthusiasts. An hour later, we were dropped off at Doncaster station, and a quick check of the timetable showed that I could squeeze in a trip to Healey Mills shed, just outside of Wakefield, before the works opened. I took a local multiple unit to Wakefield Westgate and used this journey to thumb through the Ian Allen Shed Directory for directions to Healey Mills shed.

On arrival at Wakefield Westgate I caught a Yorkshire Traction bus outside the station which would take me to Horbury Bridge on the outskirts of Wakefield. Fortunately for me, there was a friendly British Rail driver on the bus who was also heading for the shed and we struck up conversation. When we got off the bus, not only did he show me the way to the shed, but he gave me a personal tour around the complex. This was brilliant, especially as Healey Mills had a reputation for being a little unwelcoming to spotters. Sadly, though, the visit was a bit disappointing. There were only twenty-one locos on shed, although in the refueller at the far end of the yard, was my last Knottingley class 47. The BR driver turned out to be a real star that day, even going so far as to arrange a lift back to Westgate for me in one of the yellow BR minibuses that regularly shuttled between the shed and the station.

Back at Doncaster, the crowds were building, and a queue had formed on the footbridge leading to the Works. Murmurs of excitement grew as we anticipated which locos might be in the workshops. As usual, there was the token 'know-it-all' in possession of a TOPS listing who was giving audience to whoever would listen. This was something I never understood. Why would you want to know what was in the works when you were literally about to go and find out? The unknown, for me, was part of the experience.

After what felt like an age the gates opened, and the crowds moved off the platforms and into the works. Doncaster, like most major workshops, provided a good benchmark for the state of the locomotive fleet overall. At one extreme, in the back of the works yard, were 31012 and 24066 waiting patiently to be dismantled, whilst, in another part of the site, the finishing touches were being applied to the latest of the new class 56s being constructed at the plant, 56047 and 56048.

Having checked out the works and browsed through the merchandise stalls, the next step was to take the well-trodden path to Doncaster Carr shed, just to the south of the station. By now, it was early afternoon, and there was a steady procession of people walking down the dual carriageway leading from the station to the shed. When I arrived at the depot, though, things didn't look too good. A large crowd was blocking the access road and car park area, and the foreman and a couple of charge hands were doing their best to keep order. Essentially they were trying to prevent people from entering the depot but were rapidly losing the battle. It didn't help that one member of a railway society had an official group permit for several enthusiasts which allowed access to the shed. And 'several', it would appear, was not quantified!

A selection of Open Day related tickets.

Chaos ensued as British Rail staff tried to establish who, out of the hundred or so people milling around the depot entrance, were with the railway society and therefore potentially allowed to enter.

At this point, I noticed one of the older guys from our coach sizing up the situation, so I started chatting to him about the likelihood of getting into the depot. After a moment, he retrieved an orange safety vest from his pocket and told me to follow him. We approached the foreman, who appeared highly stressed.

"Looks like you've got a problem here, mate," stated my new friend. "Tell you what, I work on the railway so if you like, I'll follow the official party around at the back and make sure nobody tries to tag along. Provided," he gestured at me, "you let my lad join the group."

The foreman agreed, relaxing somewhat, thankful perhaps for the small gesture of support and an extra pair of hands. For me, well, there couldn't have been a better outcome or end to the day.

Doncaster shed had bags of character and oozed history. Thirty-seven locos were present, including a brand new 56046 along with shunters 08065 and 08066, both of which had been withdrawn from Immingham the previous September and were waiting for movement to the works to be dismantled later in the year.

This wasn't the first time I'd been to the depot, though. The previous year, on a Bristol excursion to York (which just happened to be on a very hot Sunday in June), I had taken the opportunity to nip down to Doncaster and see what I could find. I had walked out to the works and managed to get a few numbers by accessing the lane running between the backs of houses and the perimeter of the works (known by enthusiasts as 'dog sh*t' alley). After this, I'd headed to the shed yet, as I approached the entrance I'd noticed some significant-looking offices between me and the depot as well as more than a few '**TRESPASSERS WILL BE PROSECUTED**' signs.

Steeling my resolve, I'd continued down the path to the shed, carefully avoiding eye-contact with a couple of official looking people I met on the way before finally reaching the back of the shed. This had been much quieter and, adrenalin pumping, I had walked up and down the lines of resting locos, grateful not to have been caught.

On my latter visit during Doncaster Open Day accompanied by my high-vis wearing buddy, I'd marvelled at the shed's structural size and lack of significant change since it had held its steam allocation. In later years, the roof would go on to be removed and then the site rationalised before eventual demolition.

The trip back to Bristol from the Doncaster Open Day was uneventful with even the coach driver seeming more alert. I guess if you are stuck in Doncaster for ten hours and have no interest in railways, then sleep is a great option!

As a teenager, money was always tight, so I learned to make use of any travel offers British Rail promoted. In the summer of 1978, one of the best offers was the *"Three Return Trips"* special, which allowed you to take three trips in a single week to a range of destinations ranging from the South West up into the East Midlands and the North West. The price at under £10.00 was a bargain; the only drawback was that the stations from where each trip had to start were all in South Wales.

The previous week, I had spent with Mum and Dad in a holiday home at Ringwood on the edge of the New Forest. This had proved to be a good base for getting in visits to Bournemouth Branksome depot and Eastleigh. In fact, I made sure I was at Eastleigh Works on Tuesday lunchtime so as not to miss the weekly tour of the works. On this particular week it produced twelve locos - including a Stratford-based shunter under overhaul, which I needed.

The 'Three Trip Bargain' that I did in 1978 was the first 'longer' trip I had done. The ticket alongside doesn't tell the whole story because Brian and I engineered a fourth trip by avoiding too many 'grips'. The ticket below the 'Three Return Trips' was from my first trip to Scotland. Note the price of £4.00. You would struggle to get a sandwich in the buffet for that now!

The week in the New Forest was rounded off by Dad stopping at Westbury on the way home, where a quick sprint around the small depot produced another twelve locos. An ex-works Crompton, 33019, was the main attraction there.

The following day, Sunday, 27th August, was the beginning of the three-trip special (offer), which Brian and I had decided we would do. I went to Temple Meads before Brian that day to buy a blue Inter City shoulder bag. These were popular at the time but unfortunately had a design flaw - the buckle holding the strap to the bag was of such poor quality that anything heavier than a paperback resulted in the bag becoming detached from the strap! This bag, purchased in anticipation of our three-trip adventures, was actually the second one I had bought.

Brian and I arrived at Temple Meads at 23:00, which meant a long wait before taking our first train to South Wales where we could start our three trip offer. The train wasn't due until 01:14, though luckily for us, the station was busy with empty stock movements in preparation for Bank Holiday extra services (we were travelling on the August bank holiday weekend). Our wait, therefore, was not in vain and we saw Stratford-based 47150 on shed sporting its unique Stratford silver roof.

Eventually, the 01:14 Temple Meads to Swansea (hauled by 47069) arrived and Brian and I boarded as far as Newport - this being our chosen starting destination. Unfortunately, due to timings, we were faced with a six-hour wait before our next train, the 07:53 service from Newport to Birmingham, so we set up camp in the heated waiting area. This was a dusty room full of the ghosts of travellers past and furnished sparsely with some rickety old chairs and a large wooden table. We stared at the oversized ticking clock; it was going to be a long night.

Why, Brian and I wondered as we waited, did we always seem to start extensive trips with an overnight stop or long journey? Often a sound idea at the planning stage, it always backfired when it came to the actual trip. This, though, was not our first time waiting so we employed our resourcefulness and began playing 'our' version of the 1978 World Cup Football shove penny competition – the rules for which are well known amongst those with a lot of time on their hands.

Equipment is minimal. All you need are three two-pence coins and a reasonably large rectangular table. The object of the game is to flick one coin through the gap created by the other two coins and so on until two of the coins are within an inch of the end of the table that you are attacking. The third coin is flicked through the gap to score a goal. If you miss the gap or hit one of the other coins, the possession passes to your opponent.

Three hours and many games later, we were ready for the final. It was to be Brazil against Bristol Rovers. Whilst this draw may have raised a few eyebrows, it didn't to us, though even we were surprised when such a game was decided by a single Paul Randall (my favourite Rovers player) goal three minutes from time. The (1978 Shove Penny) World Cup was coming to Eastville!

As the hours rolled on, we looked for the first signs of daylight in the night sky for we planned to walk out to Ebbw Junction shed as soon as it was light enough. At around 05:00, a faint light appeared from the east, so we left the waiting room and started the long trek to Ebbw Junction, unsurprised to find the streets very quiet - Newport was asleep.

We made good progress, but it was still dark when we got to the shed. Neither of us realised how long it would take for the sun to rise, however we were there now, so we entered the shed and managed to stir a dozing foreman who looked a bit startled to be awoken. Fair play, he gave us a torchlight tour of the depot. Sadly, we didn't see much, just seven 08s, fourteen 37s and a 47. We said farewell to our host, allowing him to return to his slumbers(!), and made our way back to Newport station, which was starting to come to life. Brian and I were already feeling tired, though. A stark contrast to the other fresh-faced travellers! At last, the 07:53 Birmingham service (45018) pulled into the station, and the real trip finally began. Apart from noting a pair of 20s on Gloucester shed, the journey was uneventful from a spotting perspective then, when we arrived at Birmingham New Street, we changed to a Manchester train (86213) and continued our journey north.

The plan was to spend the day visiting Manchester sheds. There were three main depots: Longsight was a converted steam shed, Reddish was a purpose-built depot for the electric motive power operating on the line between Manchester and Sheffield, whilst Newton Heath was a new depot built on the site of a large steam shed and was primarily a multiple-unit depot with a small loco servicing shed attached.

After pulling in at (Manchester) Piccadilly station, we alighted then began our walk across the city centre towards the then 'down at heel' but still magnificent looking Victoria station.

From here, we took a local service out to Dean Lane to visit Newton Heath depot. The shed was a five minute walk from the station and we were given permission to go around although here weren't many locos other than some class 08 shunters, a few 25s and lots of units.

25104 pausing at Manchester Victoria whilst the driver phones the signal box. A local DMU waits on an adjacent platform, on the 3rd of July 1980.

Leaving the shed we took a train back to Victoria and walked across the city centre back to Piccadilly. We then picked up a local service to Ashburys station before walking to Longsight shed. It seemed to be open house at the sheds, and at last, we saw our first class 40, ironically 40001, on the depot. Again, there were more multiple units than locos.

After leaving Longsight we returned to Ashburys station and caught a local service out to Guide Bridge station where a path led to the loco stabling point which was holding twenty two locos taking a Bank Holiday rest from working freight trains over the Pennines. As the day and afternoon wore on, we continued using local trains and eventually made our way to Reddish Depot.

In 1978, times were already changing at Reddish, with nine of the 1950s constructed class 76 electrics withdrawn and in various states of decay littered around the depot site. It was an incredibly sad story. The depot had initially been built solely for the maintenance of electric locos and units which worked the Manchester to Sheffield electrified line via Woodhead; however, a year after the opening of the line, a BR corporate decision was taken to move away from the 1500-volt DC system (that was uniquely used on this line) to 25K volt AC. This effectively meant that the end was coming before things even really got started. Whilst both systems use energised overhead electric cables, the 1500v system was based on tried and tested technology -which is still used in the Netherlands – however, the lower voltage (1500v as compared to 25KV) meant more electricity substations and extensive cabling were required. Initially this system had been proposed before the Second World War but the inevitable delay meant that subsequent developments in electrification technology resulted in 25K volts becoming the national standard for BR before 1500 v got off the ground. This was simply because 25KV was considered to be more efficient. All was not lost, though. Reddish, when faced with the changes to the electrification infrastructure, diversified by maintaining diesel locomotives in addition to the electrics which meant there was always an interesting selection of locos.

The first open day at Reddish depot had taken place on Sunday the 9th September 1973. The foreword to the programme booklet was written by K.J Davies, Manager for the Manchester Division, who talked excitedly about improvements scheduled for the next ten years, including High-Speed Trains, the tilting electric Advanced Passenger Train and the proposed underground railway linking Manchester Piccadilly and Victoria. The first was a success, the second a failure, and I'm not sure what happened to the Manchester underground!

It was getting towards the end of our first full day on our 'bargain' trip, and the effect of having no sleep was beginning to take its toll. The plan, though, was to keep going with an overnight vigil planned at Crewe before heading to Birmingham and then home the next morning.

Exhausted but still in good spirits, Brian and I spent eight hours at Crewe from 20:00 through to 04:00 and amassed exactly one hundred loco numbers. This included through traffic as well as movements on and off the busy diesel depot just to the south of the station.

Reddish Open Day Programme from the 9th of September, 1973.

Crewe at night was a magical place with an almost constant procession of passenger, freight and parcels traffic. Pioneer electrics from classes 81 through to 85 (nicknamed 'roarers' due to the sound of air passing through their cooling fans) all made an appearance together with 86s and 87s. We also saw another eight class 40s, including 40076, which did not move all week, whilst 20069 was an unusual visitor, presumably waiting to enter the works.

On the train back down to Birmingham New Street, we drifted in and out of sleep, arriving there just after 05:00, weary and tired. New Street was busy with many DMU movements in preparation for the rush hour, and the subterranean air was thick with exhaust fumes.

Though tired, we still had a bit left in the tank so Brian and I took the first train of the day out to Bescot, then rounded things off with a visit to Saltley depot before boarding the 09:15 off New Street back to Bristol. This was hauled by 47338 and, with the exception of a Knottingley-based 47 stabled on a freight at Gloucester, that was the only thing of note.

The design of the three-trip special ticket meant it had three separate sections, one for each day. The trick was to try and avoid getting the ticket punched on the last (third day) section of the ticket – in the hope we could squeeze in an extra day. Arriving home after our first day, I crawled into bed at 11:00 a.m., which felt like coming home from a night shift, mainly as we were due to head back to South Wales again later that day to start our second trip.

A few hours of sleep later, it was back to the grind and a late evening bus down to Temple Meads. We were planning to take a very early morning train from Bristol up to Birmingham to kick off the second day but as we did not want to wait around at Temple Meads twiddling our thumbs for a few hours we decided to do a quick round trip to Cardiff and back. The short trip out was fine however we had problems on the return service from Cardiff back to Bristol. The train had been stoned by Cardiff City fans near Canton on the approach to Cardiff. The Cardiff fans were targeting Spurs fans returning from a League Cup match at Swansea which meant a long delay whilst a replacement train was put into service. This ended up being a silver lining for us with the new train consisting of a few old Mark 1 carriages hauled by 37180, an engine that would more commonly be found on freight work.

The second trip started in earnest with the 01:10 from Bristol up to Birmingham (as did so many railway adventures from Bristol then), which would be followed by another long wait at New Street before making our way up the West Coast Main Line to Stafford where we changed for a service to Manchester. Our chilly early morning wait at Stafford was significantly brightened when 40121 powered through the mist with a northbound tanker train.

We decided to kill some time in Manchester before heading to Crewe, where we had a permit for a 14:00 visit to the works so Brian and I walked from Manchester Piccadilly to Victoria once more before heading out to Newton Heath shed for another visit. There wasn't as much on shed as there had been earlier in the week, but at least we saw another two 40s.

Arriving at Crewe in good time for our 14:00 visit, Brian and I then made our way out to the works. Crewe was still a significant railway town in the late 1970s, and the works were essentially unchanged since steam days. It still employed nearly five thousand people and occupied a considerable tract of land on the north side of the Chester line.

Built in the 1840s, Crewe works purpose was to construct and repair locomotives and stock for the Grand Junction Railway. Ownership later passed to the London North Western Railway and eventually the London, Midland and Scottish Railway before it finally became part of British Rail Engineering Ltd. Crewe works is surrounded by streets with rows of terraced houses - many of which would have been occupied by railway workers - which creates a strong sense of railway history. Brian and I arrived to join the motley crew of spotters waiting by the main gate for our guide to appear, and after a few minutes (following a cursory inspection of permits), we were allowed in.

Though it may seem unlikely, the railway scene is constantly changing, and amongst the seventy-odd numbers, we noted several withdrawn and rusting class 40s whilst, in the erecting shop, new HSTs E43100 through to E43105 were nearing completion. These HSTs would go on to start a forty-year (and counting) stint, thrashing up and down our main lines. This, the second of the three-day trip, ended at a more leisurely pace, and we finished with a journey from Crewe to Chester, travelling in a DMU that had broken down and had been rescued by 24047 - one of the few remaining class 24s in service. A total of one hundred and fifty-one class 24s were built at Derby, Crewe and Darlington workshops as one of the pilot diesel classes but due to the rationalisation of the BR fleet, the class was phased out from the 1960s, and by the late 1970s, only a handful, all based at Crewe, were still operating.

A short walk from the station took us to Chester depot, which was primarily a multiple unit shed, but there would usually be few locos present as well. This visit produced another two of the increasingly rare class 24s. Our return trip from Chester to Crewe was just as good as we were hauled by 24082. The rest of what turned out to be a warm summer's evening was spent on the busy platforms at Crewe, where every now and again, the sounds of train movements through the station were drowned out by roars from Gresty Road where Crewe Alexandra FC were beating Notts County 2 – 1 in the League Cup. We wound the day up by catching a service back to Bristol arriving at Temple Meads just before midnight.

The next day (in effect our third trip on the ticket), we had a morning off and caught up on some sleep before taking the 15:08 Bristol to Derby service. Our plan was to squeeze in a quick trip to Toton. Of note on the journey was 56036 in its new large logo livery at Burton and former Southern Region electro-diesel 74010 on Etches Park depot. The 74 had been moved to Derby for possible use as a mobile generator following withdrawal the previous year; however, the project was abandoned, and the loco was eventually scrapped at Doncaster in 1979.

On arrival at Derby Midland, we headed for the bus station to catch a Barton bus service to Sandiacre from where we left the bus and walked up to the grass bank overlooking Toton Depot and yard. A two-hour stint here produced nearly forty numbers, including two of the pioneer 'Peaks', 44002 Helvellyn and 44007 Ingleborough.

It was a lovely, warm summer evening on Toton bank and we got talking to a couple of the local spotters. With the conversation flowing, Brian and I both wished we'd had a tent to camp there for the night but keeping to our plan meant we needed to take another Barton service on to Nottingham for a quick visit to the station before catching a late-night train back to Derby.

After an hour of waiting at Derby, we caught the 00:28 service to Birmingham, which was hauled by 45069. Unfortunately, the 45 failed at Burton, resulting in 45115 being called out to complete the journey delivering us to Birmingham New Street an hour late. Luckily, I had fallen into a heavy sleep at Burton, so many of these antics passed virtually unnoticed.

Heading into the final (fourth) day of this mini adventure (which was the extra day on the three-trip ticket due to careful use and avoiding getting the last day punched by the ticket inspector), we repeated Monday's itinerary by taking the first service off New Street out to Bescot (which was again busy with over forty locos noted). We followed this with a visit to Saltley shed and then back to New Street.

The focus on this last day was going to be the Liverpool area, so we took the first available train up to Crewe where we then boarded a local service to Liverpool which stopped at Ditton, a few miles from Liverpool. From here, Brian and I walked out to Widnes West Deviation Yard to pick off the Allerton-based shunters working there. This proved to be a bit of a trek through a grim industrial landscape, but it was successful and gave us another class 40 as well as the shunters.

The walk back was equally as long and it didn't help when we ended up with a bit of a wait at Ditton before boarding the next service to Liverpool. From this Liverpool service, we alighted at Allerton and then took the short walk from Allerton station along Woolton Road, which led us to the depot.

Allerton depot opened in the late 1960s and was a key London Midland Region depot. Initially, it had an allocation of class 40s and 25s, although, for much of its life, it only held an allocation of shunters that worked throughout the Liverpool area along with a significant number of diesel multiple units. When Brian and I arrived at Allerton depot, we paid our customary visit to the foreman's office, gave our charity donation and were then given the all clear to walk around the shed. Allerton's five-lane through shed and stabling sidings had a reasonable amount on view for a weekday, although apart from seven shunters together with a class 81 and an 85 electric, the rest were multiple units. A fifteen-minute walk took us from Allerton shed to the sidings at Speke - the large marshalling yard to the south of Allerton – and here we could see a mix of locos, with another class 24, 24081, being the stand out attraction. From Speke, we then trekked back to Allerton station and caught a local service to Edge Hill, where we again walked out towards the shed.

Liverpool has much in common with Bristol; both are major ports with a problematic history and gained considerable wealth from their involvement with the slave trade. In more enlightened times, I think it's fair to say that both cities have struggled to come to terms with some of the less salubrious elements of their heritage.

Chasing railway engines can often take you to the less attractive parts of cities and on this trip, Brian and I noticed that Liverpool seemed to have more than its fair share of poverty and social issues. The closer we got to Edge Hill shed, the more apparent this became. Going down Tiverton Street, for example, we observed a long row of run-down terraced houses, some boarded up, with kids running around, appearing happy and scruffy. A far cry from the more comfortable existence I was fortunate to inhabit.

A view of Birkenhead Mollington Street depot on the 30th of June 1980.

Back in the days of steam, Edge Hill had been a big shed, and the present small two-lane diesel servicing shed occupied only a fraction of its previous site. On that day we saw only a couple of locos so Brian and I headed back to Edge Hill Station for the short run down through the impressive cutting into Lime Street. Once at Lime Street we swapped trains for a low-level service out to Birkenhead. The low-level services to Birkenhead were operated by class 502 and 503 Electric Multiple Units (EMUs), which ran between Liverpool, the Wirral and Southport. These EMUs were nearly forty years old and were introduced in the same year as WW2 started. I'm not sure if this says more about the longevity of these units or about how much was being invested in rolling stock at that time.

The short journey under the Mersey to Birkenhead Central took only a few minutes where we were able to see extra units in the shed next to the station. From there it was only a five-minute walk along Argyle Street before entering Mollington Street, a short cul-de-sac which led to the depot. Another former steam shed this was now home to new occupiers and Brian and I were given a friendly welcome. We were able to explore the old, run-down shed at our leisure though there wasn't much to see except for four class 40s.

The next stop was Birkenhead Electric Depot, just outside of Birkenhead North Station. Again, we were fortunate to be allowed around, and there were many more units to see, but unfortunately, the 08 shunter usually based there was absent.

The following few hours were spent filling our books with the antique electric units we noted travelling around the area, plus visiting Hall Road depot along the way. It was coming to the end of our trip by now, and Brian and I were both weary. When we boarded a late-night train to Crewe from Lime Street, I immediately fell into a deep sleep leaving the guard at Crewe needing to wake me up due to the train terminating there. Whilst the guard's actions were perfectly reasonable, my brain had not fully connected and though I usually consider myself mild-mannered, on this occasion I felt it necessary to launch a torrent of (undeserved) abuse at said guard. I am blaming this out-of-character behaviour on exhaustion, and though it may be a bit late, I would like to extend my sincere apologies to that guard.

The return journey to Bristol was, in the same vein, a bit of a sleepy blur, but as we arrived at Temple Meads very early on Saturday morning, we faced an additional wait for the first train to Sea Mills before we could trudge off home. This did have its benefits, though, as we were able to note 47094 rumbling through the station, hauling china clay empties back to Cornwall.

1981 was welcomed in at the Mill House pub in Stoke Bishop (my second closest pub to home) with some friends. A few days later, on 3rd January, Brian and I headed north towards Birmingham, hauled behind 47502. We were on a trip combining railways and football because it was the third round of the FA Cup, and Bristol Rovers were playing Preston North End away at Deepdale. Rovers had only won one game all season, so the likelihood of progressing further in the cup was slim, yet we were in high spirits as we boarded the train on a dull, overcast day.

At Birmingham, 86326 took over from the Duff (nickname for a 47) for the next leg of our journey to Wigan. Once here, we planned to walk out to Springs Branch shed, which looked quite full as we passed by. Alighting at Wigan North Western, with our shed directory in hand, it took us about twenty minutes to reach Springs Branch shed by heading off down Warrington Road and then turning right into the little cul-de-sac which led down to the depot. In the early 1980s, Springs Branch was still a busy depot and there were thirty-three engines on shed, including thirteen class 40s. Springs Branch was a mix of the old and the new, with a purpose-built three-lane diesel depot complimenting the former steam shed which was now used for stabling.

After our visit to Springs Branch, we returned to Wigan North Western and took 86225 for the short run up to Preston. Having some spare time, we took advantage and called in at the small stabling point at Ladywell, just to the north of the station where 40001 was resting for the weekend. From there, it was a twenty-minute walk to Deepdale where Brian and I were surprised by the lack of people heading to the ground. The game, it seemed, had failed to excite the Preston football faithful. A paltry 6,348 brave souls turned up for the match that grey afternoon which was a far cry from the massive crowds following PNE (Preston North End) in their glory days in the First Division.

At half-time, things were going our way. Rovers were leading 4–0, and we were in Gas heaven, along with the hundred or so Rovers fans who had made the long trek to Lancashire. The only problem was that some disgruntled Preston supporters were trying to climb over the fencing to get into the away end in order to demonstrate how upset they were.

As the second half got underway, PNE started to fight back and pulled the score back to 4–3. Luckily for Rovers, the referee then ended the match; another five minutes of play would definitely have cost us the game.

Despite the spirited comeback, Preston supporters were still unhappy and, being heavily outnumbered by fans from the home side, we were not looking forward to the walk back to the station. As it turned out, police and stewards held us 'away' fans back from leaving the ground for some time after the game to allow the home fans to clear. Shortly thereafter, the police pulled back the large, corrugated iron gate which had kept us in - and then simply disappeared into the shadows! This was not ideal as several Preston supporters had decided not to go home immediately and were waiting outside wanting a fight. Brian and I made our way across the car park as quickly as we could whilst mayhem broke out with boots and fists flying everywhere. In the end we almost ran back to the station and continued to glance nervously behind - just in case.

Rovers didn't get beyond the next round of the cup (away to Southampton -Kevin Keegan et al.), so after all that drama we saw yet another cup run grind to a halt during a season that would eventually see us relegated.

The journey back to Bristol from Preston started with 87018 down to Crewe, 86321 to Birmingham, 47502 (again) to Cheltenham and then 50041 Bulwark for the final run down to Bristol. Four classic engines, of which only 47502 (now privately owned by Harry Needle of the Harry Needle Railroad Company) managed to survive the eventual cutters torch. Reflecting now, it saddens me how we used to take being hauled by engines such as these for granted back in the 1980s.

That trip to Preston to watch Rovers became a regular fixture for a couple of years and often involved visiting some engine sheds on the way up. On the way home, one particular trip ended badly when we were caught in a snow drift between Gloucester and Bristol. The train had to stop for a couple of hours and because it was so cold, ice formed inside the compartment. At one point, I thought Brian had lost consciousness, so I began to shake him, desperately concerned for his welfare. Brian was, of course, just sleeping and was less than receptive to my assistance!

As the years went on it became increasingly apparent that the seemingly harmless and relaxing hobby of trainspotting, where a couple of hours could be passed away at Temple Meads jotting down numbers, had transformed into a hardcore activity which could push you to your limits. Typically, now, trips were lasting for a few days, and often involved catching a couple of hours sleep here and there, whether that be on a train, or in a waiting room or anywhere really. Anywhere that wasn't a proper bed.

At times, I found this to be both physically and mentally exhausting, and I remember, as I write this, making use of several late-night or overnight services, just to make the excursions more manageable. One such trip I recall, involved catching a late-night train from York to Hull, which was made longer by a few hours break at Doncaster. This fell in the middle part of a three-day slog around the North East. By the time I arrived at Hull Paragon station, the first watery signs of daylight could be seen in the sky, and I left the station on yet another long trek. The streets were empty as the city slept, whilst I walked alone towards my goal - the shunters at Hull Docks.

Other memories of that time include trying to find a bus on a Sunday evening from Shirebrook back to Chesterfield to catch the last train back to Bristol. 'Bus' and 'Sunday evening' were never a good combination!

After long overnight trips Brian and I would regularly be like zombies on the early morning commuter trains as we tried to stay awake and not fall into the central gangway or onto a neighbouring passenger. The 01:10 service from Bristol to Newcastle, was often the starting point of these expeditions and it was usual for these overnight services to attract more diverse travelling companions than those you might typically find during the day. An aggressive squaddie on leave from Plymouth was one that sticks in the mind who was constantly blasting out Boston albums on his huge 'ghetto blaster'.

Sometimes it was necessary to try and find a different compartment in the old Mark 1 stock in an attempt to avoid some of the more colourful fellow travellers. On the 01:10 train, though, this could be a bit of a challenge as majority of the train comprised parcel vans, so there were only a couple of passenger coaches to choose from.

The Big Bike Ride : June 1981

In the late 1970s, the 250cc Honda Superdream was the UK's best-selling motorcycle. It was a sleek, seductive and eye-catching machine (well, it was to me anyway) which proved popular with teenage bikers but generally disliked by motorcycling purists for its unconventional looks, slow performance and the fact that it sounded more like a sewing machine than a motorbike - all of which were largely true.

My friend Mike introduced me to motorbikes in 1979, and his silver Superdream - which he kept in immaculate condition - was his pride and joy.

I had just left the sixth form at Bristol Cathedral School, and to mark the end of school life, a group of us booked a caravan for the week at Littlesea Caravan Park just outside of Weymouth. Mike and his mate Steve, who owned a Suzuki X7 250cc bike (more conventional than the Superdream and significantly quicker due to its 2-stroke engine), came down on their bikes while the rest of us travelled by train.

At the time, I had not really been interested in bikes; however, one morning during the holiday, Steve took me on the back of his X7 along the coast road out to Osmington and back. Steve was the archetypal gentle giant - well over six foot in his leathers and with a strong West Country accent, but he rode his bike with little regard for his (or anyone else's) safety.

All I can say about the bike ride was if there had been any speed cameras along Weymouth front, then I don't think they would have been quick enough to pick us up as we screamed out towards Osmington. Worryingly, Steve only gave me two pieces of advice before we set off: 'not to put my feet down' and to 'lean when he did', which both sounded reasonable.

On that short journey, though, my understanding of the laws of gravity and physics significantly increased. I had not realised (as we banked and dipped into corners), that it was possible for two people on a bike to get so close to the ground without parting company from said bike. Nor had I appreciated that it was possible to squeeze through such a small gap down the middle of the road at high speed in the face of oncoming traffic.

After arriving safely back in Weymouth, I got off Steve's bike with my legs like jelly and pure adrenalin coursing through my body. I made an instant decision. I needed to get a bike.

I started off with a 50cc Suzuki moped (this was my dad's and had been getting dusty in the garage for a few years due to him now having a car) and then, after a while, I upgraded to a brand-new Honda 125 twin.

Having started working for Bristol City Council at Avonmouth Docks (which at that time was owned and operated by the Council) I was able to afford the new Honda and then, the following year, I part-exchanged it for my very own 250cc Honda Superdream.

Incredible. This felt like a major achievement. I could officially call myself a biker.

Not only that, but I was able to prove a point to my mates. Being of a slighter build than most, some of them took great pleasure in telling me that the Superdream would be too big a bike for me to handle. Oh, how I enjoyed showing them they were wrong!

Bike magazines were now purchased each month alongside my staple railway magazines and I soon discovered that my new and old hobbies complimented each other perfectly. I had an exhilarating means of transport to travel around the country, which made visiting sheds, especially those off the beaten track, that much easier. On a good day, I could combine a bike ride, a few loco sheds and a Rovers away game. I was living the dream.

Seeing my new-found freedom meant it was only a matter of time before Brian acquired his own bike, a red Honda CB200 with a throaty two-into-one exhaust which made the bike sound far more powerful than it was. It was also guaranteed to raise a few eyebrows when passing through built-up areas.

I think, when Brian bought his bike, that was the moment I finally realised how badly I had caught the motorbike bug and that shockingly, it was almost on a par with trainspotting.

True to type I had grown my hair and bought the obligatory leather jacket from Bristol's rambling Eastville Market, which was held every Sunday in the Rovers ground car park. The market was huge and a big draw for people all over Bristol. It was the place to get quality ("honest my luvver") leather jackets at a good price. Though it was not strictly a bohemian market, Eastville always attracted a good cross-section of people, including punks who could pick up alternative gear for a much lower price than at Paradise Garage by the Broadmead bus station in Bristol's main shopping area – their other 'go to' retail preference.

I already liked bands such as Motorhead, Iron Maiden and Status Quo, and most weeks I would go to 'Heavy Rock Nite' at Tiffany's (long since demolished and now the site of a BUPA hospital). Or I would watch start-up bands such as Def Leppard (they made it) and Vardis (they didn't) at the famous Granary club just off Queen Square in central Bristol.

As our confidence in riding bikes grew, so the trips that Brian and I took became more ambitious and inevitably, I suppose, we started planning the 'Big Bike Ride'. We were to spend ten days in June 1981 visiting as many depots as possible.

In deference to our interests at the time, most of the planning for this trip was done at the Lochiel, a floating pub moored opposite the Arnolfini in Bristol's old dockland. The Lochiel, built in 1939, had spent most of her sea-faring days delivering mail for the David MacBrayne ferry company to the Western Islands off the coast of Scotland. She had faithfully plied her trade for over thirty years before being purchased by the Bristol-based Courage Brewery and converted into a floating restaurant.

Brian and I became regulars in the Cargo Bar (which was located under the forward deck), and every Sunday, a disco would be held on the little dance floor adjacent to the bar. The pop music scene was dominated by Adam and the Ants at the time, so we'd have a great laugh watching people doing their own version of the unusual 'ant music' dance, which generally seemed to involve strutting around and pumping your chest out whilst moving your arms in a flamboyant way.

As ever, money was an issue. Neither of us had much for our trip, and we knew we wanted to take both bikes. After a bit of head-scratching, it was agreed that the best way to keep costs down was by taking tents and camping overnight(s) before moving on the following day. Thus, the final two weeks before our 'Big Bike Ride' were spent buying camping equipment and getting the bikes ready.

The night before we were due to leave, Brian and I rode down to the Lochiel for a last pre-holiday drink. During the evening, we met up with a couple of girls and thought we would try and impress them with a ride on the back of our bikes around the dockside. Unfortunately, in the excitement of the moment, I forgot to take the padlock off the front disc of my bike (the steering lock had broken a few years earlier, so I'd resorted to padlocking the front wheel for security), and, as I began to pull away with my new found friend on the back, the padlock hit the front forks causing the bike to come to a sudden halt.

My passenger rapidly lost interest which was (worryingly) of less concern to me than the fact I had managed to shear through the speedometer cable.

Sadly I didn't learn from this mistake. A few years later, I did exactly the same thing but with a heavier bike - a Honda CX500. The story was similar. I'd met a girl in a Bristol club a few days earlier who had invited me round to her place. After a couple of hours, I left and thought I would dazzle her by roaring off on my bike. This time, though, my omission in removing the front wheel padlock had a more catastrophic outcome. The padlock almost instantly jammed the wheel and threw me off, resulting in the (heavier) bike pinning me to the ground. Fortunately, the girl I was trying to make an impact on had a friend with her, and between them, they managed to lift the bike off me. Needless to say, that relationship did not flourish either!

Anyway, back to the big bike ride. Having made appropriate repairs to my Superdream, on the afternoon of Friday, 19th June 1981, Brian and I negotiated our two overloaded Hondas down the road towards our first planned stop at Bath Road Depot. We'd wanted to call in at Bath Road on our way out of Bristol (force of habit if you like) to see what was visible from the bridge overlooking the yard, but as soon as we pulled into the car park, a high-vis vested official came rushing out and told us to 'sod off'. Not the best of starts!

The plan after Bath Road was to head for Brighton via Westbury, Salisbury and Eastleigh, so a while later, under leaden and overcast skies, we pulled into the car park at Westbury Depot, thankfully to a much warmer reception. Sadly, we didn't see anything here we hadn't expected to: a couple of out-stationed Bath Road 08s, seven 47s, a 31 and a 37, so we continued our trip not long after arriving.

Following a brief stop at Salisbury, we reached Southampton and threaded our way through the commuter traffic towards Eastleigh and then eventually found the station. To reach the depot, we needed to head down Southampton Road, turn left over the railway and then into Campbell Road, where we parked the bikes. The lanes running behind the houses on either side of Campbell Road skirted both the depot and the perimeter of the works, so we craned our necks, trying to catch sight of anything of interest.

The shed itself was busy with engines coming on for the weekend therefore we didn't bother asking to go around – we knew it was unlikely that we'd be allowed and in any case, we had quite a long ride to Brighton ahead. In addition, we weren't sure if we needed to be at our chosen campsite before a certain time to be guaranteed a pitch, so we figured it was best to keep going. As we left Eastleigh depot, though, we picked off two works shunters before taking the A259 and our onward journey towards Brighton.

The ride took longer than expected, and we arrived at the council-run campsite in darkness, which made putting the tents up for the first time an interesting challenge. Neither of us had thought to pack something as useful as a torch and this was way before mobile phones or any technology that is now helpfully equipped with this beneficial tool.

As with most campsites, there was a café/bar offering food and hospitality. Brian and I were both starving so we entered gratefully – only to discover we had missed the food ordering deadline by five minutes. Our protests fell on deaf ears, and we had no joy whatsoever from the 'jobsworth' charged with catering arrangements. Thus, our first glorious day ended with a few pints and a selection of Golden Wonder's finest crisps, which was to become, perhaps not surprisingly, our staple diet for the next ten days.

Saturday 20 June 1981

After a restless night trying to get comfy in a sleeping bag on hard ground, we woke to the sound of seagulls and a bright sunny morning. Heaven. This was definitely the life. Until I opened the tent flap and realised our campsite was overlooked by a council rubbish tip and the seagulls were scavenging for scraps which ruined the beauty of the moment somewhat.

Brian and I set about packing up the tents and sleeping bags – which took infinitely longer than expected. Hoping we would get quicker at this task as the trip went on, we loaded up our bikes and set off for the next leg of our journey.

The plan for the day was to head for Hastings, Dover, Ashford, Tonbridge and then finish off at Redhill, where I had bagged a bed for the night from my Auntie Sylvia, who I had not seen for many years.

We made a brief stop at Brighton station and noted a pair of Hither Green based 33s double heading a service on the Hove platforms, whilst a 47 was stabled in the depot yard with three 73s and an 09 pilot, before heading out on the A259 towards Newhaven. Here we were hoping to find the 08 shunter that worked the yard there. Having spent half an hour trying to locate it, we eventually realised it was stabled directly underneath the road bridge just outside the station - but not until we had driven over the bridge several times! Though we didn't know it at that point, we would go on to spend much of this trip trying to find shunters that remained somewhat elusive.

After Newhaven, we took the coastal road through Seaford and on towards Hastings, where the weather decided to take a turn for the worst. Biking across Romney Marshes in the mist and rain was less than fun, so we were feeling cold and particularly miserable when, out of the foggy greyness, appeared a roadside burger van. Salvation! Twenty minutes later, full of strong, sweet coffee and greasy cheeseburgers, we continued on to Hastings.

St. Leonards Depot is located almost on the beach about a mile west of the centre of Hastings on the north side of the Bexhill line. It consists of two sheds at either end of a yard; one building was used for maintenance, whilst the other was a running shed for multiple units. The main interest for Brian and I were the class 33 'Slim Jims' (33201 – 33212), but none were visible outside of the shed, so we decided to try our luck and ventured into the maintenance building. Nobody seemed bothered by our presence, so we walked through the building, noting down a few class 202 and 203 slim-line diesel multiple units as we went. The only loco, 33209, was found on jacks at the far end of the depot.

Before returning to our bikes we spent a while skimming pebbles on the beach and then set off for Dover just as the sun began to break through, bathing the road in watery sunshine. Much like our journey the previous night, the ride to England's busiest ferry port also took longer than expected, and it was nearly 1.00 pm before we found our way to the stabling point by Dover Marine Station. Here, we ticked off three Hither Green 33s and a pair of Ashford 08s, although annoyingly, there was another shunter partially hidden behind wagons in the Town Yard, which we were unable to see.

By now, it had really warmed up, so we gave the bikes a chance to cool down and took a stroll along the pier adjacent to Dover Marine Station, taking care not to trip over the sea angler's rods. The last thing we wanted to do was incur their wrath, and the hooks (which flashed past us as they cast out their lines) looked rather dangerous. Neither of us fancied having one of those catching on our clothing – or worse! The walk enabled us to view the yard alongside Marine station where another Hither Green 33 was stabled along with a few multiple units. We had the luxury of seeing these from the pier which was a first for us both, and made me wonder how many other places you could trainspot at such a picturesque setting.

Leaving the coast behind, we headed inland for the short run to Ashford, where we found the station easily enough and left our bikes in the car park while we went to the stabling point. Here, we saw a few more 73s and a 33 laid up for the weekend, along with a couple of units.

The next stop was Ashford Depot, though we had little confidence in being able to see anything as the locos tended to be stabled in the servicing shed sandwiched between the main line and the workshops, which were difficult to access without a permit. After a bit of scouting around, however, we did manage to find a couple more shunters.

It was approaching the end of our second day with just two more stops planned – one to pick up the shunter at Paddock Wood and, lastly, the stabling point in Tonbridge yard. The ride along the A262 towards Sissinghurst was excellent; the road was fast and dry, and the afternoon sun was warm, but not so hot as to make the ride uncomfortable. Arriving at Paddock Wood in great spirits, we saw the shunter and a 73 powering through on a Dover-bound ferry van train.

The final part of our day we had, when planning this trip, considered to be straightforward. The locos at Tonbridge are stabled in the yard to the west of the station, and we knew there was a footbridge which literally passed over the roofs of the stabled engines. Try as we might, however, we could not find our way to the footbridge and somehow ended up in a summer school fête! From here, we tried to pick off the numbers using our binoculars, but that didn't work, and we were starting to attract concerned looks from several of the fête attendees. Worried we might end up being arrested, Brian and I agreed it was time to move on.

More searching ensued and eventually we found the yard and the footbridge, from which we were rewarded with a small selection of locos. Bizarrely, we must have passed the steps leading to the footbridge several times but simply not seen them.

At last, Redhill loomed which is where we were due to stay with Auntie Sylvia and Uncle Pat for the night. Even though I hadn't seen them for several years, they made Brian and I incredibly welcome, and provided us with plenty of food and comfortable lodgings. It was great to sleep on a proper bed, although having only camped for one night, I knew we'd need to get used to sleeping in the tent as there were many nights under canvas ahead.

Before we left, I checked the brakes on my Honda – for some reason, they never worked as well as they should, always feeling spongy and lacking real 'bite'. Despite this and endless subsequent attempts to sort them out, the problem with the brakes was never resolved (though the one thing I never changed was the brake disc, so it could well have been that). They worked, which was the main thing so I resolved to try being a less 'fussy' biker as I never seemed happy with the condition of my bike.

Sunday 21 June 1981

Wishing Aunt Sylvia and Uncle Pat farewell, we left Redhill and headed off up the A23 towards London. I felt on top of the world. Everything was rosy, and I even gave a cheery wave to the police speed gunner at the side of the road as we passed by (probably a little too quickly for our own good).

On this, our second full day, the first planned stop was Selhurst, where luckily the foreman had gone walkabout, and the charge hand let us into the shed. Here, we noted the shunters stabled for the weekend together with a solitary 47 before moving on to Norwood Junction which was quiet as anything - most of the locos were presumably out on engineering work.

The next stop at Hither Green produced a predictable collection of 33s and 73s, with our visit to New Cross Gate immediately afterwards proving disappointing (it was empty) and then noting only one shunter at Bricklayers Arms parcels depot.

Our original plan after these morning visits had been to go through the Blackwall Tunnel and head off to East Anglia; however, as it was only 11.00 am, we decided to do a bonus mini-London circuit. This took us to the sheds and stabling points at Marylebone, Willesden, Old Oak Common, Stonebridge Park, Finsbury Park and Cricklewood, all of which we were able to get around except for Cricklewood, which was something of a building site. The footbridge leading to the shed from the car park had been demolished, and the new electric multiple unit shed to the north of the depot was being used for stock stabling.

As morning progressed to afternoon, we left London on a day when the temperature was steadily rising and followed the A12 to Colchester. Though we were revelling in the arrival of summer, the necessary protection of crash helmet and leather biking gear made the journey rather warm and a little uncomfortable. Initially, we had planned to visit Stratford, but, finding ourselves pressed for time, we skipped this, knowing that we would be visiting its Open Day the following month. The A12 was mostly dual carriageway, which made the journey somewhat monotonous, and the hour it took for us to ride to Colchester felt more like three. Eventually, though, we parked our bikes in the station car park and made the short walk to the depot.

Old Oak Common was always a popular London shed to visit with its unique turntable; a remnant from the former steam shed. We visited the depot on Sunday the 21st of June 1981 as part of our bike tour. 47189 and others were around the turntable.

The small two-lane through shed at Colchester is just to the south of the station on the west side of the line and is a typical 1960s Eastern Region construction. We were granted access to go around and noted a fair spread of locos though none of the local 03 shunters. The only unusual visitor was 37184 from Landore, the sight of which would have made a nice change for any local spotters, but not for us.

Once we'd finished, the foreman suggested hanging on for a bit as there was a return Clacton excursion due through with a Bristol Bath Road 47 on, but after consideration, Brian and I decided to give it a miss. We did appreciate the thought, but I don't think he realised we were from Bristol. Besides, it was now close to 6.00 pm, and we still had a few stops to go.

Parkeston Quay was next but proved disappointing as there were about ten locos down in the yard but they were too far away to get the numbers. This inspired a last-minute decision to head to Harwich, which was more worthwhile and produced another local 08 working the ferry sidings, and a cop for me.

Our last planned stop of the day was at Ipswich, where a grumpy foreman let us go around the stabling point. Here, we found twelve locos on shed, including two 03s. Though it had been a long day, Brian and I had covered eleven depots/stabling points and ridden across London, so now it was just a final short blast up the A12 to a campsite at Saxmundham. Unfortunately, we arrived just after they stopped serving food (again!) and again, they would not budge, so it was back to the crisps and lager and our less-than-comfortable sleeping quarters.

Monday 22 June 1981

It's great waking up under canvas on a warm sunny morning with the heat filtering through the tent accentuating the smell of grass and tent fabric. Our rough plan for today was to cover Norwich, March and Peterborough in an East Anglian arc and, with the weather fine, we packed up quicker than after our Brighton stop and were back on the A12 again in no time at all.

This is my loaded Honda Super Dream 250 at a campsite in Saxmundham, two days into our big bike ride. We visited just under one hundred depots and stabling points during this two week break in June 1981.

Arriving in Lowestoft, we picked off the 08 shunter in the yard at the west end of Lowestoft station before making our way to the seafront for breakfast (coffee and cakes!), then headed further up the coast to Great Yarmouth. A Norwich-based class 03 was stabled in Vauxhall carriage sidings just outside of the station whilst a train heating unit, DB968013, was in the station. This heating unit had started life as a class 31, then following its withdrawal from service in March 1979, had been converted, along with three others, into units that were used to pre-heat carriages before use. I had first seen it as a class 31 at Stratford depot a couple of years earlier on the way to Norwich. Sadly, its life as a heating unit would not last long though, and a couple of years after our trip it was scrapped at Doncaster.

The temperature was really picking up now as we headed to Norwich, with the tarmac shimmering beautifully in the heat haze. The huge Norfolk sky spread endlessly above us, and with both bikes running smoothly, everything in the world was great. Norwich, though, turned out to be frustrating as we could not get around the depot courtesy of a very stroppy foreman. Worse, my last Norwich 03 was somewhere in the depot but not visible from outside.

Norwich had a large allocation of these veteran shunters that were built in the late 1950s and living in Bristol made it difficult to see the ones based in East Anglia unless you made a specific trip. And even having made a specific trip, it would seem there was no guarantee. Sadly, Brian and I admitted defeat and acknowledged that it was time to move on.

The ride from Norwich to Kings Lynn was long and hot and the road was straight and fast, which was great, but I wished the bikes had a bit more grunt as both lacked the power to really do the road justice. A late lunch came courtesy of an American-style roadside diner, which looked strangely appropriate in a landscape of endless sun-drenched farmland.

Full up and rested, we jumped back on the bikes and re-joined the road, the sun hot on our backs. It wasn't until a few miles later that I became aware of a car behind me flashing its lights and getting a little too close for comfort. I increased my speed, but the car was still on my tail and continued flashing its lights. As Brian was out of sight behind me, I decided to pull over and find out what the problem was, just in case something had happened to my friend. The car duly stopped behind me and the driver got out – a big bloke who appeared quite lairy. I watched cautiously as he approached.

"Your tent fell off your bike miles back; I was trying to catch your attention," he said.

How silly did I feel? Having built this up into something it clearly wasn't, I thanked the man profusely and then waited a few minutes for Brian to arrive. But when he did, he wasn't alone. He was in possession of my wayward tent, which he had stopped to retrieve from the roadside.

A few years later I had a similar experience whilst riding my Honda CX500 on another railway themed trip to East Anglia. I was heading up the A1 heading towards Stevenage when, out of the corner of my eye, I noticed one of my hard pannier luggage cases become detached from the bike. I could see it bouncing down the road and watched in horror as it hit the front of a limousine that had been behind me. The road was busy, which luckily meant that speeds were on the slow side. The limousine and I pulled over, and a chauffeur exited the luxury car. Collecting my pannier case from the roadside, he brought it back to me.

"I think this is yours, sir," he said, oh so politely.

I nodded a little awkwardly. "Sorry about that. Did I do any damage?"

"It appears to have smashed the number plate; £10 should cover it," he replied.

Gratefully, I handed over a much smaller sum of cash than I had initially feared and carried on my way – concerned nonetheless at my worrying habit of losing objects off my bike.

Back in 1981, after what seemed an age, we rode into Swaffham and stopped at a roadside pub to get out of the hot sun and quench our thirst. Taking our helmets off, Brian and I walked into the bar to be met with silence from the five or six elders nursing their pints. After sitting down with our drinks, the locals resumed their conversation, and within moments, the rich Norfolk accent drew us in. We listened intently to all the local gossip, which was dominated by births and bereavements, before moving on once more. At Kings Lynn shed, we picked up another two 03s before heading off down the A47 to March.

There has been an engine shed at March since 1850 and its reason for being, was primarily to provide locos to work freights out of the adjacent Whitemoor yard which, when it was expanded in the early 1930s, was one of the largest yards of its type in the country. The diesel depot opened in 1963 and was built close to the site of the former steam shed. It was a standard Eastern Region concrete and glass structure with three lanes at each end and a central maintenance section, plus an adjoining two lane servicing area. Sadly, we had no joy getting around, but being such an open site, we saw almost everything anyway.

After a long hot ride across Norfolk, we reached Kings Lynn where 03154 was stabled in the station. I wish I'd remembered to move our bike helmets before I took this shot! The date was the 22nd of June 1981.

31325 stabled at Kings Lynn between duties on the 22nd of June 1981.

Leaving the shed behind, we made our way to the station where the nearby coal yard was full of withdrawn Class 306 electric units. These were waiting for the final tow to A. King and Sons at nearby Snailwell, where they would be scrapped. These first-generation units were initially built in 1949 before being rebuilt in 1960 and spent most of their life working out of Liverpool Street on the Shenfield line.

After March, we headed for Peterborough, where a long line of shunters were conveniently lined up outside the small single-lane servicing shed just north of the station. We spent a short time at the station where I saw my second from last class 31 running light engine, going through the station heading south. I would have to wait another six years before seeing my last one, Thornaby-based 31141, ironically on Bath Road depot. It was always a bit of a personal achievement to see all the engines in a particular class, especially ones like the 31s, of which the majority of the two hundred and sixty-three members were predominantly based away from Bristol.

We finished the day at one of the TRAX campsites (a group of campsites linked to horse racing courses and good value for money) at Southwell, not far from Nottingham. We had trouble finding this one, though, and didn't get the tents set up until quite late. As was the 'theme' of our trip, we began to panic that we wouldn't make the bar before closing time and, as it had been a long hot day, the attraction of a cold pint of lager was rapidly increasing.

Luckily, we reached the bar just in time and ordered a few pints each to keep us going - how well they went down! The sudden rush of alcohol made it a bit of a stumble back to the tents, but once we made it, sleep came quickly!

Tuesday 23 June 1981

The good weather was still with us the next morning as we packed up and headed to Lincoln. The plan was to visit Immingham, Hull and York then see how far north we could get before we needed to call it a day. The campsite guidebook told us there was another TRAX site next to Thirsk racecourse, so we decided to aim for that.

Our first stop was Lincoln Central Station, where we left our bikes for the short walk to the depot. We entered the main building and headed for the foreman's office, where we were given permission to go around. There, we picked up three 08s and a good selection of the class 114 Derby-built two-car units that had been based at the depot for a few years. The depot itself had been built in the late 1950s – primarily for the servicing and maintenance of DMUs, although it also had an allocation of both class 03 and 08 shunters together with, for a time, some class 31s.

The depot would only last another six years after our visit, sadly, and closed in October 1987. It has been given a new lease of life now, though, and currently operates as a bus depot.

When we'd seen everything at the depot, we returned to the station to get some breakfast, however we were greeted with a torrent of abuse from the car park attendant. It was mainly directed at me when I went to retrieve something from my bike, but we had no clue what the issue was. Bemused, we departed the car park, resigned to waiting a little longer for breakfast.

Before going on to Immingham we tried to find the shunters working at Grimsby West Marsh Yard and Grimsby station, but neither of them were at their locations. I should say that in these pre-internet days, we, like many others, relied on publications such as 'Shunter Duties' produced annually by the Inter City Railway Society, which listed all the engine sheds and the associated duties of the shunters allocated there. This was how we knew where to go, and usually, they were pretty accurate.

Immingham was a major depot and boasted a large allocation of engines throughout much of its life. In the 1970s, it had a big allocation of 47s, which tended to operate freight traffic associated with the docks and neighbouring industrial hinterland, so it was always a treat at Bristol to see Immingham 47s drafted in to haul additional summer Saturday services to the West Country.

Immingham depot was a mix of the old and the new. The former Great Central steam shed was used for stabling locos (and loco lifting) whilst servicing and general maintenance were carried out in another typical Eastern Region 1960s steel and glass constructed depot. This had a four-lane maintenance section at one end and a three-lane servicing section at the other, with a stores area in the middle. Dominating the site, though, was a huge concrete coaling tower, which could be seen for miles due to the flat landscape. This was a relic from the steam age and lasted almost as long as the diesel depot, which eventually closed in the early 2000s and has now been demolished.

After parking our bikes, we went to say hello to the foreman which involved walking across the front of the servicing shed. This enabled us to see a fair number of engines before being told to go away – which almost inevitably seemed to happen here. We tried to get into the docks for a better view of the far side of the depot and almost persuaded the security chap at the gate to let us in, but it was not to be when he called his boss for advice. Still, we didn't mind too much as we managed to see another six locos in Immingham West Yard before heading for Scunthorpe. On the way, we stopped at Barnetby, where we parked the bikes and then hiked down a cinder path to get the shunter at Wrawby Junction yard. It was a long, hot slog in our bike clobber, and I encountered problems with a vicious dog who kept snapping at my ankles. All of that to see 08514, which wasn't even a cop!

Our pub lunch was taken just outside of Scunthorpe before going on to visit Frodingham Depot, which was another standard Eastern Region three-lane dead-end structure, similar in design to many others. After Frodingham, we spent two hours at Scunthorpe station itself, which was busy, producing twenty-five locos hauling a steady stream of steel, coal, oil and ore traffic. Leaving Scunthorpe, we decided to drop in at Goole to pick up the shunters, although it took over an hour to eventually find them. We'd somehow missed the Hull road at Goole, but when we finally reached Hessle Yard (just outside of Hull) we saw the shunters, then went down to the docks to see the shunters working there. After that, we visited Botanic Gardens shed, located a mile or so from Hull Paragon station on a short spur from the Hull to Goole line.

Botanic Gardens was a shed I had never managed to get around and this day was no different - although we did see most of the locos. It had originally been intended as a DMU depot but did house a few shunters to work the local yards and the docks. There were often some mainline locos stabled as well, particularly on the weekend. It is one of the few sheds that survives to this day, albeit in a much-reduced form. There are only two lanes left of the original building, and the adjoining yards have been converted into football pitches.

Time was now against us, and the ride from Hull to York took a good while. The sun started to slip below the horizon as we travelled which made a perhaps tedious and lengthy ride very scenic. The evening light bathed the countryside in a golden glow and I was reminded that this was what motorcycling was all about. Being out on the open roads with the sweet smells and sounds of summer, hanging in the air.

By the time we reached York, the light was poor, but we managed quick visits to the shed and station before battling up the A19 to Thirsk, where we decided to call it a day. I'm not sure why, but in Thirsk we encountered a 'no bikers allowed' rule at three separate pubs in our quest for food and drink, so we had to admit defeat. Clearly, something had happened previously but that did nothing to help Brian and I who were by now hungry, thirsty and more than a little deflated.

In the end it was decided for Brian to go to the campsite and put up the tents whilst I went in search of an off-licence for some takeaway beers.

Wednesday 24 June 1981

Our weather luck ran out and we woke to heavy rain. Damp and cold, the ride from Thirsk to Darlington was miserable and more than once I cursed the disappearing summer. Luckily, we were able to breakfast at Darlington station which improved our moods somewhat, before going on to the depot just north of the station. Though we could not get around on this occasion, we did manage to scramble up a grass bank and see a few numbers, which was better than nothing.

Darlington shed was purpose-built for DMUs and opened in 1957. Like Hull, it had an allocation of shunters with usually a few visiting mainline locos. Though built for diesels, Darlington looked like a standard Eastern steam shed in 1981. Today, sadly, there is no trace left whatsoever.

Unfortunately, the weather did not let up and was no better at Thornaby, which can be a pretty grim place even on the best of sunny days. Thornaby depot opened in 1958 and was the last shed built for steam. It was also part of the new Tees yard complex and was converted to maintain diesels in the late 1950s after only a short time in use as a steam shed. We couldn't get around the depot, which was not entirely unexpected as this was a busy freight base with a lot of engine movements on a weekday; however, we did manage to see about twenty locos from the outside. I'd first been to Thornaby a couple of years earlier on a hot weekday in August and was fortunate to be allowed around. On that day, there were about thirty engines on, and for someone from Bristol, the Thornaby-based 31s and 37s that I noted were like gold dust as they rarely ventured down to the West Country.

From Thornaby, Brian and I headed on up the north-east coast. By the time we reached the outskirts of Sunderland (via Middlesbrough Goods yard and West Hartlepool stabling point), we were soaking wet which necessitated a detour into a pub. This we found on a run-down looking 1970s housing estate where we enjoyed a cheap and cheerful egg and chips lunch. Refreshed and slightly drier, we eventually found Sunderland docks where one of my last 37s was pulling out of the yard on a coal working.

Sunderland shed was one of my favourites. Accessed from a lane off Prospect Row (how great a name is that for a street in a dockland wilderness), it was a throwback to the steam age where nothing had changed apart from the new order of shunters and class 37s lurking in the dim shadows. Dull and grimy, these locos were part of the north-east scene for many years and spent their life hauling coal, steel and chemicals, just like their sisters down at Canton.

From the depot, we went onwards to Sunderland station, where two Thornaby 31s passed through. Then it was up to Newcastle with a pause on the way at Felling NCL Depot to see the local class 08 shunter. The next stop was Gateshead shed on the other side of the Tyne from Newcastle Central station, though again, we could not get around.

Gateshead (we discovered) had quite a strict policy on visits and was one of the more difficult sheds to access, although Brian and I still managed to see sixteen engines by making the abortive trip and trying to reason with the foreman.

I always considered that Gateshead was to Thornaby like Haymarket was to Eastfield. Gateshead was the top link depot providing the mighty Deltics for the London and Scottish services and Peaks/Brush for the cross-country diagrams, together with a smaller allocation of freight engines and shunters. Thornaby was less glamorous than Gateshead with far more freight turns. In the late 1950s, Gateshead had one of the biggest steam allocations on the network, with over two hundred engines based there. It was converted to a diesel depot in 1965 when the shed closed to steam.

Despite being unable to access the shed during our 1981 visit, I did manage to get around it once and found it to be an interesting building comprising a high-arched central roof section with a turntable in the complex. It was a solid brick-built construction with eight covered lanes and a yard at the back, which you could not see from the entrance (though it was visible from passing trains). I later became aware that you could scramble up a bank to access the yard, which is something I wish I'd known earlier! The majority of the engines were hidden in the main building so were impossible to see unless you were granted access, though there was also a two lane servicing shed between the main shed and running line where you could sometimes get a few numbers.

Tyne Yard was next, (another large freight yard a few miles south of Newcastle), where we spent twenty minutes on the long road bridge that crosses the north end of the yard. We watched a steady stream of freights rumbling by but unfortunately the small depot in the yard was a bit too far away to be able to see much, despite several attempts to find a favourable vantage point. From Tyne Yard we worked our way back to Newcastle Central station, where we took a break and I bought the current month's Motorcycle Mechanics magazine which had a feature on the new Yamaha XJ550. I really wished I'd had some of that power! Then it was off to Heaton, where I saw my last Gateshead 03.

The original plan to go to Blyth after Heaton and visit the depot there had to be abandoned due to lack of time; we needed to get to our planned campsite at Gretna which was a shame, because I never got a chance to visit that depot in latter years, either.

The rain finally began to ease off as the day wore on to the point where evening sun broke through, just in time for us to leave Newcastle on the A69. After about fifty miles we turned off the A69 and onto the A6071 to Gretna and then arrived at our campsite.

The site was called 'First House' which, being located in the first town in Scotland after crossing the border from England, was not only apt, but it also matched nearly every shop, pub and restaurant name. Pretty much everything was prefixed by 'First'.

Brian and I set the tents up before walking back to a hotel just outside of Gretna and close to a railway line (railway underbridge) that we had passed earlier. We stayed until closing by which time I had become quite taken with the barmaid, and though I didn't actually speak to her, deep down, I sensed she felt the same way. A side effect I think of spending too much time with Brian!

When the hotel bar eventually closed, we trudged back to the campsite. I was a little saddened to think there would be no romantic interlude on this trip with the closest thing to romance being the twinkling lights of Kingmoor Yard, which were visible in the distance. On the walk back we spent half an hour on the railway bridge and watched a few trains roll by, including our only true Scottish engine, a class 27 trundling northwards on a short freight.

Before retreating to our tents and sleeping bags, Brian and I decided this would be as far north as we would travel. The following day, we would return southwards through Cumbria and into the industrial heartlands of Lancashire and Yorkshire.

Thursday 25 June 1981

The cafe at the 'First House' campsite (complete with gingham tablecloths) provided tea, toast and Scottish marmalade which was greatly appreciated before we set off on our bikes for another day of exploring.

Our first stop was Gretna town centre where we bought 'topical' postcards to send back to workmates though we were a little unsure how they might be interpreted. The sun had once more hidden, but at least it was a dry start, and we made good time down to Carlisle Kingmoor shed, our first planned stop of the day. Here, we were met with a surprisingly friendly foreman who let us have a look around. Kingmoor was a comparatively new depot, having opened in 1968 on the opposite side of the main line to the previous steam shed of the same name. It was a good-sized four-lane through shed with stabling sidings to the south. There were fifteen locos on shed, and the foreman also gave us a list of the Kingmoor shunters along with details of where they were working. The last Kingmoor shunter that Brian wanted was at Dumfries; however I had just copped my last one whilst we were walking around the depot so we decided to split for the day. Brian made his way back over the border to Dumfries whilst I went down the west coast and skirted the Lakes via stops for railway interest at Workington and Barrow. We agreed to meet up later at Carnforth station.

Workington was quite busy with a few locos on the stabling point next to the station; however, the shunter that worked the yard to the south of the station was tricky to find. After eventually locating it, I began the ride from there to Barrow, which was slow due to a lot of holiday traffic on the road. Even being on a motorbike, I still found myself held up by tourist coaches and caravans, which were difficult to pass on the narrow roads. To make matters worse, the weather was starting to turn and by the time I reached Barrow, it was raining.

The stark industrial landscape I now encountered, which was dominated by the shipyard, was a huge contrast to the mountains and hills of the Lake District, which were still visible in the distance. The traffic and weather meant that by the time I arrived at Carnforth, Brian had already been waiting a good hour. We decided it was time to break for food and I watched enviously as Brian unwrapped his family-size Gala pie (basically a pork pie with an egg in the middle) that he had been working through for the last few days. The pie was a monster - this was the early 1980s, and family size meant just that. Brian was now in sight of the egg, so the end was near. My food by contrast, consisted of a cold meat and potato pie from a shop by the station.

Whilst we were eating, a class 86 electric rattled through the station on the main line to London, making the station buildings shake as it passed. At that time, Carnforth station was run down; the buildings were dilapidated and needed some care, which I always thought was incredibly sad given its provenance; Carnforth having been the location for one of the most romantic and iconic railway scenes in British film history - none other than the location where Celia Johnson and Trevor Howard experienced their 'brief encounter' in the world-renowned film of the same name. By the late 1980s (when Brian and I visited), it was a shadow of its former self, but fast forward twenty years and the station buildings have now been fully restored and are an international tourist attraction.

After lunch, it was time to get on the A6 and head south to Preston despite the inclement weather. We incorporated a quick visit to the stabling point just north of the station at Ladywell before arriving at Wigan Springs Branch shed. At long last, the sun broke through and having been given the okay from the foreman, we went around the depot, happy to be both there, and enjoying a break from the leaden skies. For a weekday, we were surprised to find over twenty locos on, which was more than Brian and I expected. I liked Springs Branch; it was a good traditional London Midland Region diesel depot.

The last stop at the end of a long day was Warrington Bank Quay, where another twenty locos were chalked up either on the stabling point or busying themselves in Arpley yard. Once we'd finished at Warrington we headed for Acton Bridge where we found a bike-welcoming campsite. Not only was everyone very friendly, just over the swing bridge on the nearby Weaver Navigation waterway was *The Leigh Arms*, which was a great-looking pub with food and music. Brian and I were set and looking forward to a good night!

Music at *The Leigh Arms* was provided by a local folk group who worked through several covers during the course of the evening, including the 1968 classic from Mary Hopkin, 'Those Were the Days'. At this point - whether it was due to the long day on the bike or the alcohol we were enjoying - I found the song really resonating with me emotionally. This led to a sobering thought process.

The song, as you might know, is about being young and carefree but is tempered by reflection when you are older and thinking about freer times which have now passed by. I was twenty years old at the time and had spent much of the last ten years being obsessed by railways. Admittedly, I loved it and got so much pleasure from it (and still do to some extent), but for the first time that night, I had a real sense of getting older and the passage of time from youth to adulthood. I could not see myself doing this for another ten years, and with regard to my job, I wasn't at that time heading for a fulfilling career. I had also not really considered the possibility of settling down in a long-term relationship.

Not wanting to dampen the mood, I kept these thoughts to myself, yet even now, some thirty-five years later, I can still remember the depths of sombre emotion I reached that night.

It's strange how songs can have such an impact on you. When I was thirteen, I was overcome with sadness whenever I heard Terry Jack's 'Seasons in the Sun'. I think that was the first time I had a sense of the impact of death, and the lyrics had a profound effect.

<u>Friday 26th June 1981</u>

Anyway, enough of this melancholy.

The next morning, my sense of maudlin had left, thankfully, and we had another busy day ahead of us. Today would be less travelling and more visits, with a plan to take in Liverpool and Manchester and end the day at a campsite just outside Buxton. So, accompanied by a light drizzle, we loaded up the bikes for a pre-breakfast visit and then began our journey to find Northwich depot.

Northwich was another excellent old vintage London Midland steam shed. It comprised a four-road red brick dead-end shed, unchanged from the day the last steam engine had been hauled away to the scrap yard. The history of this depot can be traced back to the 1870s when it was constructed by the West Cheshire Railway Company before being absorbed and upgraded by the London Midland Region in the late 1940s.

Two class 40s were on shed along with a couple of 25s. There should also have been two Allerton-based 08s in the area, but we only managed to see one. Today, Northwich depot has been replaced by a housing estate, but it managed to survive into the late 1980s before being demolished.

Getting back on our bikes, we headed off to Chester, though not before enjoying a good fry-up at a roadside cafe just outside the city. We were the only customers, and the multi-tasking owner/chef/waiter fussed over us in a way that we both agreed we could get used to!

It was a struggle to find Chester depot, and when we finally got there, we were surprised the foreman allowed us to look around as there were a lot of train movements. The depot consisted of two separate three-lane former steam sheds and whilst primarily a multiple-unit depot, you could usually see a few locos. We noted three shunters and a couple of 25s, as well as lots of units. Today, the old depot buildings have been replaced by a modern structure which continues to service multiple units in the area.

After Chester, we had another change of plan. Brian wanted to get his last Chester shunter, which the shed foreman said was working in Croes Newyd yard just south of the Wrexham stations, so off we went and after an hour's ride, we found it. Brian was happy, though we were now running behind time as we headed up towards the Wirral.

Hungry, we stopped at a pub in Tranmere for some lunch. No one could say we weren't seeing the gritty side of England on this trip, as we were yet again faced with what can only be described as a bleak industrial landscape. Names such as Port Sunlight which can be found in this area, seem oddly congruent. Okay, so I know Port Sunlight was a model village designed by the Lever Brothers, and the name relates to one of its popular brands, but even so, it seems ironic.

Thornaby based 47302 on Chester Depot on the 22nd of November 1986.

Nicely filled, we continued up the Wirral peninsula to Birkenhead diesel depot, one of my favourites, with its soot-covered walls, leaking roof and throwbacks to a bygone age. Birkenhead had previously been a big steam shed consisting of two separate eight-lane sheds adjacent to each other; one was for the Great Western Railway, and the other for the London and North Western Railway. In later years, a two-lane diesel maintenance shed was built on the side, initially to look after a batch of Hudswell Clarke shunters which arrived in the 1950s to be used in the docks.

Birkenhead seemed to be struggling to let go of its steam shed roots and was interesting to go around. Some of the tracks had been lifted, but we could sense how busy it must have been back in the day. As it was a weekday, the shed was quiet, with only a couple of class 40s and 47s on.

The drizzle persisted as we continued our mini tour of the Wirral peninsula. Times were changing almost before our eyes, and when we reached the electric depot at Birkenhead North, one of the new class 507 electric multiple units was on shed, surrounded by its soon-to-be-replaced 1940 elder relatives.

Further signs of change were found in Birkenhead Docks, where another five of the ancient units were waiting for the tow to the scrap yard; no more trips to sunny Southport for them.

Heading for Liverpool via the Mersey tunnel, we began a slow trek to Warrington with various stops along the way. First was Lime Street station followed by Allerton shed, which required another charity donation (and lapel badge), to gain access in order to be rewarded with seven 08s. Then it was onto Speke Yard, Garston Freightliner Terminal (two more 08s) and then Warrington Arpley stabling point, where we notched up another three class 40s.

Time was getting increasingly tight, especially as we struggled to find Newton Heath (our next stop) and lost valuable time riding around north Manchester. We then lost even more time writing down the sixty-plus diesel multiple unit numbers; still, with ten locos on shed, including another four 40s, things were on the up.

After dropping into Victoria, we crossed town to Longsight, which was also stacked out with another seventy units and an impressive twenty-one locos (nine 40s), although three of them, 40016 (ex Campania), 40023 (ex Lancastria) and 40031 (ex Sylvania) had all been withdrawn a few months earlier and were dumped in the yard at the back of the shed.

It was now approaching dusk as we reached Reddish Depot, and there was a touch of deja vu about this visit as nothing much had changed since my last visit three years earlier. The same rusting 76s were littered around the site; broken cab windows replaced by wooden boards, paintwork faded, and one, 76056, just a body shell, resting between the shed and the Fallowfield loop line. It looked like a portent of doom to the remaining class members as the Woodhead electrified route across the Pennines linking Manchester and Sheffield was to close the following month. The rest of the class were subsequently withdrawn and cut up over the next few years, mainly at CF Booth's scrapyard in Rotherham.

Despite this, Reddish depot lingered on for a while, servicing the Class 506 electric units, which continued to operate on the Manchester – Glossop – Hadfield part of the system. Not long after our visit, this line was also converted to 25kV, which marked the end of Reddish depot; it ended up closing for good in April 1983. Today, the site has been redeveloped with housing, which feels like an ignominious end to a depot which once maintained the Midland Blue Pullman sets. If you're reading this and live in Reddish on Falshaw Way, Salwick Way, or Fylde Lane, maybe spare a thought for the rich railway heritage upon which your homes are built.

76032 with pantographs raised at Reddish on the 3rd of July 1980. A year later, the depot would close.

Due to time pressures, we gave our next planned stop, Guide Bridge, a miss and headed straight for our campsite at Buxton. Previous experience had taught us that the kitchens at campsites tended to close early (!), and some even stopped people arriving after a certain time.

Riding down the A6 to Buxton was a long slog along a busy road and inevitably, by the time we arrived at the campsite, it was dark and cold, and once again, we had missed the food.

Brian put the tents up while I went to find a chip shop.

Saturday 27th June 1981

Today was due to be another busy one as we were aiming for a circular tour of the South Yorkshire sheds, ending up at the Sheffield University campus and an overnight stop with Al, an old schoolmate, who was studying Architecture there. The day started bright, but it was cold, and I couldn't get warm. The tents were damp and a pain to pack up, but it was a relief knowing we wouldn't need to sleep in them that night. We would be in the comfort and warmth of Al's digs.

The first stop was Buxton Depot, located just off the end of the station. It was a two-lane dead-end shed typical of London Midland-designed diesel depots of the late 1950s and usually held a mix of units and a few locos which were used on the stone workings from the quarries in the area. There were five engines on shed that day, all from different classes; 08, 25, 40, 45 and a 47. Buxton closed in 1997, with its activities being moved to the stabling point and train crew depot at Peak Forest, which had been established a few years earlier. Buxton depot was finally demolished in 2016.

Still trying to warm up, we took our bikes across the Pennines to Chesterfield on what should have been a highly enjoyable ride: good surface and lots of bends, but to be honest, breakfast and warmth were the main motivations for some quick riding. We got to Chesterfield around mid-morning to find the station buffet closed, but we did manage to find a 'Little Chef' (a popular roadside café in the 1980s/1990s) on the way to Barrow Hill, where at last we were able to warm up with tea and crumpets (these being cheaper than toast as advised by the helpful staff).

Barrow Hill is one of my favourite sheds. It was a down-at-heel old roundhouse steeped in history with a working turntable which I had previously visited a few years earlier on a snowy February Saturday afternoon. This old Midland shed, which dated back to the previous century, was not always welcoming for rail enthusiasts with plenty of 'No Trespassing' signs displayed. However, as was often the case, once you got chatting with the foreman, they usually became more accommodating, and today was no exception as he was happy to let us wander around.

Nearly twenty-five locos were on shed, including three class 56s and eleven class 20s. One of the more interesting duties at Barrow Hill was the class 03 duty at the nearby Staveley Iron Works. Two of these veteran shunters were allocated here for this work and one of them was on shed when we visited. The depot became run-down in the late 1980s and eventually closed in February 1991 after which it was heavily vandalised, However like a phoenix rising from the flames it was purchased by Chesterfield Borough Council and handed over to the Barrow Hill Engine Society. It is now a thriving heritage attraction.

The next planned stop was Tinsley, so it was back onto our bikes with a short diversion into Sheffield Midland, where we picked up the class 08 station pilots.

20035 and others stabled for the weekend at Barrow Hill depot on the 6th of June 1987.

Tinsley was one of the biggest sheds in the country at the time, perched above the reception lines to the marshalling yard of the same name. It became fully operational in April 1964 following the transfer of work from Sheffield Darnall shed.

Six lanes were at each end of the shed, with workshops and ancillary facilities in the centre. Each lane could hold two main line engines, a sure sign that some learning had been taken from the construction of Finsbury Park depot (where each lane held three engines, which often resulted in engines being trapped in the shed). There were also extensive stabling sidings at Tinsley, together with a smaller servicing shed down in the yard.

Tinsley was a shed where you could see most of the locos without having to enter the site; a convenient embankment provided a viewing gallery to the whole of one side of the shed, whilst other locos could be picked off from Wood Lane, opposite the depot.

Today was good, with over seventy locos visible around the shed and yards. The class 56s were being rolled out, and there were four on shed, with the highest number being 56092, which had been released from Doncaster Works earlier in the month. Twenty-three class 20s and two visiting class 40s were also there. Predictably, we also saw the three class 13 master and slave shunters (not a very politically correct name even if it was the 1980s!) that were created at Darlington in 1965. They had been constructed by combining six class 08 shunters into three pairs of permanently coupled units for working in the marshalling yard.

After Tinsley (where we'd lingered longer than planned), it was off to the stabling point at Masborough sidings just south of the now-closed Rotherham Masborough station. We then took a circuitous route, allowing us to pick up the shunter at Rotherham Road NCB sidings and a single class 20 at Mexborough, eventually leading us to Wath Depot. It is worth briefly mentioning the ingenious design of Wath depot. Built in an area of significant subsidence, the construction involved hinged inspection pits and separate frames, which allowed for land movement. In fact, not long after opening, the weight of stabled engines caused the floor to sink; however, due to its construction, this did not present a problem. The floor was simply jacked up, and the gap filled with spoil from a local colliery.

Tinsley depot viewed from the famous bank on the 28th of March 1987.

Wath opened as an electric depot in 1951 and was located adjacent to Wath yard, representing the other end of the electrified line from Manchester via Woodhead. In later years, the overhead lines serving the shed were removed, leaving just a couple of electrified lines in the yard. This was the key location where the changeover from diesel to electric traction took place, and there was usually a mix of class 76 electrics and diesels on shed.

The depot was an outstation from Tinsley, where nearly all the diesels present were based. Despite our best efforts, we were not having any luck getting around the sheds, having been refused both here and at Tinsley. The site, however, was very open, so we were able to note some thirty numbers. I don't know if our trip was coinciding with a traditional industry seasonal shutdown, but it was noticeable that nearly all the depots were fuller than compared to previous visits.

Though we had no way of knowing, this trip was to be something of an Indian summer, considering the fate that was to befall this region in the following few years. The large Manvers Main Colliery dominated this locality back then and formed part of an industrial landscape of coke plants, steel mills and foundries, leaving a sulphurous smell hanging in the air. But there would soon be an economic downturn and change in Government policy which would severely impact this area. Some two years after our visit, Wath shed closed its doors for the last time, and today there is no trace at all of the railway infrastructure and much of the surrounding heavy industry has also gone. Arguably, the landscaped lakes and residential developments are an improvement, but I can't help feeling that something has been lost along the way, even if the air quality is better.

The next stop was Healey Mills, another large shed in a marshalling yard. Earlier in the book, I mentioned my first visit here and the fortuitous encounter with a driver heading for the shed who kindly took me around; however, in line with our recent luck, there was no repeat. Again, though, we were able to see nearly forty locos from different viewing areas, which was more than we had expected. One of the frustrations at Healey Mills was that the foreman would often let you see a list of what was on the depot, which, in some ways, was the ultimate tease. This was especially true today as locked away from sight in the shed was 56093, the highest released class 56 so far.

From Healey Mills, we moved onto Bradford Hammerton Street depot, which was another of my favourites, an old steam shed constructed in the 19th century and later converted to maintain diesel multiple units and a few local shunters. By the 1980s, the main part of the shed had fallen into such disrepair that it didn't even have a roof; however, what it lacked in structure was compensated for with bags of character. It was a gloomy place, which simply added to the experience and there was never much to see other than a couple of shunters and some units, but it was always enjoyable.

After leaving Hammerton Street we decided to do a mini tour of Bradford and visit the relatively new Bradford Interchange, which had been built on the site of the former Exchange station. Then it was onto the old and architecturally impressive Forster Square station on the other side of the city centre before finishing the little circuit with Bradford Valley goods yard, where we ticked off the local class 08s.

On arrival at our next stop, Holbeck, we were further frustrated at not being allowed around, though fortunately, there wasn't much to see anyway. The final visit of our day was planned for Knottingley, where, at last, we were allowed around, which was just as well because the locos were grouped tightly together: fourteen class 56s and four class 08s.

By now, we were running a bit late to meet Al in Sheffield and in our haste, managed to lose each other on the city's outskirts. A quick call on the mobile would have sorted this minor hitch, but unfortunately, this technology was still a few years away for us. Luckily, we had planned for such eventualities, and a phone call back home from a good old-fashioned urine-smelling red phone box helped us reconnect at Sheffield Midland station.

With both of us now where we should be, we met up with Al, who had been a great friend for many years. It was good to see him again, especially as he always had a big smile on his face. We dumped our stuff in his room on the campus and then headed to his girlfriend's flat, a short walk away. Al and Shelly had met in their school years in Bristol, and Shelly had moved to Sheffield to be with him. After some much-needed food, we went to a student party and then to Al's. At the time, I never regretted not going to Uni, but as I got older, I did feel that I had missed out on something. In all honesty, I never felt confident enough to leave home and go to university, but having said that, I am a firm believer that you make the best of whatever situation you are in and that with the right mindset and attitude, you can achieve the life you want.

Sunday 28th June 1981

Sunday morning dawned bright and warm, and a slight hangover was swiftly dispatched by an unhealthy fried breakfast in the university campus canteen. The day was set up nicely so just before we rode off, I decided to adjust the valve clearances on my bike, which had become quite 'tappy' over the last couple of days. This was a straightforward job: take off the tank, remove the cam cover (it was an overhead camshaft), adjust the valves, replace the cam cover - job done. All was going well and Al, being a bit handy with tools, had decided to help. We were just finishing off - all that remained was for Al to tighten one of the two cam cover bolts – when said bolt, promptly sheared off.

There was a deathly silence.

We looked at each other, minds whirring, trying to think of a solution, but in reality, there wasn't much we could do. No shops or garages were open (it was a Sunday in the days before Sunday trading laws were relaxed), and the only proper fix was to get the cylinder head off so that the remnant of the bolt could be drilled out and replaced. This would not be an easy job, so after much head-scratching, we decided to secure the cam cover with the one remaining bolt (being careful not to over-tighten it!) and wedge a piece of wood between the cam cover and the bottom of the frame/petrol tank. It didn't work that well, and oil quickly found its way out, covering the back of the bike, but with no other option, we packed up and headed off as best we could.

Years later, Al told me how guilty he'd felt that day!

With this unexpected drama, Brian and I missed the stabling point at Rotherwood, which had been our first planned stop, but we did get two Tinsley 08s at Beighton Yard and a class 37 on a permanent way train at Anstey. Twenty minutes later, we arrived at Worksop Yard, where we saw another five 56s before heading off to Shirebrook Depot.

Shirebrook opened in 1965 and was a standard Eastern Region depot, two-lane dead-end shed, each lane capable of holding two mainline locos. There had been a plan, a year after its opening, to double the size of the shed, although despite contracts being in place, this never happened.

The depot was located south of the then closed Shirebrook West station and, usually on a weekend, would be packed with many engines stabled in the shed area. Often, there was also a long line of locos stabled on the adjoining Warsop Main branch, curving away from the depot. Shirebrook received some fame in the late 1960s for being the home base of the prototype diesel HS4000 Kestrel, where it was used on heavy coal workings in-between trials. Kestrel was designed by Hawker Siddeley (hence the HS), who owned the Brush Company and was intended as a model for possible future British Rail use or export markets. Though it was a powerful engine, it was also heavy and not suited for passenger work, leading to it needing modification for passenger use.

During this time, it achieved some impressive performance on passenger workings before undergoing further alterations, though the engine was eventually sold and exported to Russia in 1971. Not much is known of its whereabouts thereafter, although I recall seeing photos of it rusting away in a Russian siding a few years ago.

Parking our bikes in the depot yard we managed to see most of what was there – a total of thirty-four engines on shed. Like many depots supporting declining industrial activity, its fate was uncertain with it eventually closing in 1996, fifteen years after our visit. Today the building still exists as an industrial unit.

From Shirebrook we managed to get lost in Mansfield before eventually finding our way to Westhouses stabling point. This was situated just off the midland main line a couple of miles north of Alfreton and Mansfield Parkway, at the side of a colliery branch. The semi-derelict old steam shed was currently in use as a fuelling and stabling point for Toton-based locos that worked coal traffic in the area. It was a warm Sunday afternoon when we arrived, so we enjoyed a pleasant walk along the driveway to the depot, which was nestled in the verdant Derbyshire countryside. Apart from the distant whistling of stabled class 20s, there was little noise, and the shed seemed deserted save for a few swallows darting and swooping over the roofs of the engines.

Twenty locos were on, comprising six class 56s and the rest all class 20s. Over the course of the last two days, we had seen (except for 56077) 56072 through to 56092, which made it all the more frustrating that 56093, the most recently built member of the class in traffic at that time, had been out of sight at Healey Mills. Leaving Westhouses we continued our journey south towards Nottingham and Toton depot, the biggest shed on the network.

In the rail enthusiast's world, Toton Depot's position is iconic. Located alongside the large marshalling yard of the same name (halfway between Derby and Nottingham on the primarily freight Erewash Valley line), it opened in May 1965 and at the time was described by the London Midland's publicity machine as the 'biggest (depot) of its kind in western Europe'. Built alongside the site of the former steam shed, Toton depot spanned sixteen roads and is still a mecca for spotters to this day. The shed itself existed mainly to provide engines for freight workings, taking coal from the East Midlands collieries to the power stations and was principally a freight depot with an allocation of more than two hundred locos at its peak. It was not uncommon to find a hundred engines stabled on shed at the weekend and almost as many spotters trying to jot down the numbers.

I had been to Toton a few times before and had even managed to successfully bunk the shed (get around without permission) one Sunday morning. Today, though, we were running behind schedule due to the problems with my bike, so we knew our visit would be short.

One of the biggest attractions of Toton is the high grass bank directly opposite the depot, which provides a panoramic view of the shed yard. On a good day, provided the engines were nicely spaced out, you could probably see two-thirds of what was on show via this bank. To reach it, we worked our way off the A52 into Sandiacre and found the lane that leads past several car scrap yards before ending up at the beginning of the grass bank. Leaving our bikes at the foot of the slope, we walked up the bank, and gradually, the yards came into view the higher we climbed. Loco-wise it was all predictable stuff: lines of 20s and a few 56s, although Longsight-based 40098 and Eastfield allocated 25234 were visitors that day, which made good 'bonus' sightings.

Departing Toton, we rode into Nottingham and visited the carriage sidings stabling point and then on to Coalville, where it was difficult to see loco numbers as the yard was locked up. A scramble up a grassy bank assisted to some degree, but that did involve getting stung by a thick growth of nettles, which was a definite downside!

Our last visit of the day was to Leicester Shed, just north of the station itself. The small diesel depot was the former wheel lathe of the large steam shed - which occupied the same site. This is another location where there is usually good visibility from public spaces, so again, we were able to see most of what was there without trying to get around.

It was time then to head off to our planned overnight stop at a campsite in Fazeley, not too far from Tamworth. Although it was the height of summer, it had been quite chilly in the evenings and by the time we reached the campsite, we were both cold – again. We were greeted less than hospitably with several 'no biker' signs, but we ignored these and were able to get a pitch for the night. As was par for the course, though, we had no access to food, but there was a small bar which was (thankfully) open.

Monday 29th June 1981

After a cold and slightly stressful previous day (due in no small part to my bike problem), the following morning dawned warm and sunny, which was a refreshing change. We left the campsite a little later than planned, having treated ourselves to a lie-in, before heading into Birmingham, where it was easy to find Saltley depot and even easier to see what was on. I had been to Saltley many times previously, so I knew that a trading estate had been built on the site of the former roundhouse which gave great views of the shed.

Saltley depot was small, just a three-lane servicing shed capable of holding three locos in total, and I once again observed how the number of locos seemed to vary widely at Saltley. The first time I had visited, on a weekday several years earlier, I was really pleased to get around but there had only been nine engines on shed, which didn't really seem worth the effort. On another occasion, admittedly a weekend, I counted over forty.

Today, we found a respectable twenty-six, including 40065, which was the last class 40 Brian needed in order to clear the class. Often, a few Western Region locos could be seen here having worked up from the south west on cross-country services, which would have changed to electric traction at New Street. Today 50001 Dreadnought was stabled in the yard between turns.

A general view of Saltley Depot from the adjacent trading estate, taken on what would be the last day of our big bike tour, on the 29th of June 1981.

The main depot for freight traffic in the Birmingham area was at Bescot, but we decided against riding across the city to visit as we had both been there on many occasions. It was agreed that we would concentrate on Bletchley-based shunters for the rest of the day.

After popping into Tyseley, just to the south of Birmingham city centre and surprisingly getting around (I saw my last Tyseley 08 shunter), we went off to Rugby, Northampton, Wolverton, and finally Bletchley netting five new 08s in the process. The only downside was the weather, having changed to be incredibly hot which, in full motorbike gear, was tough going. There was no pleasing us weather wise! We were either too hot or too cold, what a pair of moaners!

Northampton Bridge Street Central Materials Depot (located on a freight line southeast of the station) proved a good example of a difficult site to find. When we eventually arrived, hot and frustrated, we were able to see - in addition to the shunter - five withdrawn Class 504 electric carriages which had previously worked the Manchester to Bury services and were on their way to be scrapped by the Bird Group at Long Marston.

Due to the problem with my engine encountered the previous day, the whole of the back of my bike (and some of my leg) had become covered in oil - the temporary fix having become very temporary - so reluctantly, we decided to call a premature end to our trip and head back to Bristol. We had not made any other plans, so it seemed a sensible decision – and we were able to stop off at Oxford and Swindon en-route to note the last loco of the trip, 08583, on Swindon depot.

And that was it. The end of a memorable trip where we had seen over a thousand engines across a railway landscape that would change dramatically over the following years.

Despite the success of our trip, Brian and I were both exhausted and in part relieved to be home again. The next few days were spent fixing my bike and taking a break from railways – though a mere couple of weeks later we were both at Stratford for its open day on 11th July 1981.

Bike rides over to Severn Tunnel Junction on a Sunday afternoon had now become a regular occurrence, and even though my Superdream was low on power, it was a comfortable bike to ride. Allowing the bike to lean into the bends on the A403 as the road wound its way up from Bristol to the Severn Bridge through Aust was a great way to relax. At times, Severn Tunnel Junction was the excuse for riding the bike rather than the locos that we would find there, but both were enjoyed often and in equal measure.

Riding a bike across the Severn Bridge can be somewhat challenging, especially when a strong south-westerly is coming in off the Bristol Channel. On those occasions, I found I had to ride my bike at a bit of an angle to compensate for the wind, but as I passed each main bridge tower, there would be momentary protection from the wind, causing the bike to right itself before I would be buffeted back on an angle again. It was as exhilarating as it was scary.

One Sunday, the 6th of September 1981, saw a typical trip to the Tunnel that resulted in twenty-two locos on shed with three Crewe Diesel-based 47/3s and a 25 breaking up the local 08s and 37s. The day's gloom had already begun to turn into dusk as I headed back towards the Severn Bridge and home.

Sunday evenings in Bristol in the early 80s were (for me) dull affairs. There seemed nothing much to do, and the afternoons yawned away into the early evening. If I was on one of my trips to Severn Tunnel, then I would often break this Sunday monotony on the way home by stopping at the Severn Bridge motorway service station, which was perched high above Aust Cliff.

View on Severn Tunnel shed with 25150 and 25221 in the background with an unidentified 37. My trusty 250cc Superdream is just visible on the far left. This was on the 12th of July 1981 when at the time, bike rides to the Tunnel were a regular event.

It was always great to see some visiting 20s stabled at Severn Tunnel. Toton based 20162 and 20169 are resting for the weekend on the 4th of December 1983.

Severn Bridge services gave more than a nod to 1960s architecture and provided a range of arcade machines on the ground floor with a large cafeteria upstairs giving good views across the estuary. Most Sunday evenings would be busy with travellers breaking their journeys or people like me looking for somewhere to go to escape from Songs of Praise.

Fourteen years later, in February 1995, Severn Bridge services was to achieve notoriety when Richey Edwards, rhythm guitarist with the Manic Street Preachers, went missing. It was here that his abandoned car was found and was the last known trace of his existence. Though the full story remains a mystery, it was well documented that Richey had struggled with mental health issues, so the resulting police investigation concluded that he had taken his own life. The service station had a history as a suicide site, so you could say the conclusion was logical; however, no evidence was ever found, and several people at the time criticised the investigation. There have been a number of supposed 'sightings' of him around the world since, but the truth of what happened to him that day remains unclear.

Today, the service station is derelict and unused. Due to a reduction in traffic caused by the construction of the new Severn crossing on the M4 a few miles to the south a new, smaller service station was built a short distance from this original site.

'Freedom Of Scotland' Trip (aka: Finding Snow!)

I love snow, the way it lights up the garden at night, the cloak of silence it creates and the disruption it causes to routine activities such as going to work or popping down to the local shops. If there is a forecast of overnight snow, I will look out of the window every few minutes, checking for the first flakes.

In early 1982, Brian and I decided to get a Freedom of Scotland five-day rover, convinced we would see some dramatic snowy landscapes at that time of year. Though we didn't know it then, this trip would be our last serious spotting marathon, coming nearly eleven years after that first trip to London in 1971. We put a good deal of planning into our Scotland adventure and obtained permits for most of the Scottish sheds; the great advantage of permits north of the border was that the authorities issued them for small numbers of visitors as opposed to large groups on (which tended to be the rule elsewhere).

Friday 29th of January 1982 saw us catching the 08.15 early morning Glasgow service from Temple Meads which departed in bright sunshine without too many signs of snow. One of the aims was to see 47269 (my last 47), a Haymarket-based loco which spent most of its time north of the border. It was, therefore, a nice and unexpected surprise to find it stabled in Crewe diesel depot holding sidings on the way up to Scotland. We wondered if it was due for an overhaul at the works.

The rest of the journey was uneventful, and by the time we pulled into Glasgow Central around lunchtime, the sunshine had long since been left behind and heavy leaden skies greeted us. The first task was to book two nights' accommodation in Glasgow for the Saturday and Sunday nights so a visit to the local Tourist Office provided us with a list of hotels from which we eventually settled on Smiths Hotel. This being a small B&B in Sauchiehall Street, about a ten-minute walk from Queen Street Station. At a cost of £16 for two nights (about £70 in 2024), it seemed perfectly reasonable. The place was run by a feisty old lady who looked as if she could deal with (and probably had dealt with) every conceivable situation arising from the running of a Glaswegian B&B.

The plan for the next 24 hours was to travel to Edinburgh on Friday evening and then catch a late-night service to Perth, where we would connect with a southbound overnight train from Inverness to London. We would then leave this service at Carlisle and, after a two-hour wait, take a return service back to Edinburgh. From here, we would jump on an early Saturday morning train out to Dundee, where we had a permit to visit the shed at 11.00 am (on Saturday). Somewhere within that schedule we figured we would try and get some sleep.

At Queen Street station (a short walk from our B&B in Glasgow), we saw our first three class 27s whilst waiting for the Edinburgh service, which was hauled by 47701. The subsequent wait at Perth was three hours which was incredibly cold. The plans that had been drawn up in front of a warm fire back in Bristol did not seem quite so attractive now. It was at this point we realised our idea of making the most of overnight train services to reduce accommodation costs, may have been flawed. With little else to do except freeze, we left Perth station and found a chip shop for supper which led to a further choice: retreat to the nearest bar, or return to the station to pick up any passing late-night traffic.

Needless to say, the bar won, and we found ourselves in part of a hotel adjoining the station, which carried the widest range of whiskies I think I have ever seen. Brian and I sampled a few, enjoying the fiery golden liquid hitting the backs of our throats - though it would be a few years before I really appreciated a good single malt. That night, the 'tots' were more medicinal and did a good job of warming us up. Sometime later, the landlord (disappointingly) announced closing time, so reluctantly we made our way back to the station. After another wait, which felt like a lifetime, a class 47 eventually emerged from the gloom with the overnight London service. The train was empty, and our carriage warm and cosy so sleep came quickly. By some sixth sense, we did manage to stir though for Mossend Yard and Motherwell depot as the train continued to head south.

After alighting at Carlisle, the wait for the service back to Edinburgh was made easier by a steady procession of overnight freights hauled by 81s, 85s and 86s. A visit by a class 27 on a parcel service also broke the pattern. Sleep again won on the journey back to Edinburgh, and not even our sixth sense could wake us up for Carstairs!

The early morning service from Edinburgh to Dundee was next which comprised 26032 hauling a short rake of old Mark 1 coaches with compartments, most of which were virtually empty. Brian and I settled in, turned the heaters on full and crashed out on the cushioned bench seats on either side of the compartment. It was the closest we would get to comfort, and the compartment quickly became a warm fug, with the only inconvenience being the need to remove condensation that was restricting our view from the window.

As we passed over the Tay Bridge, a watery red sun rose through the early morning mist, and we began our approach to Dundee. On arrival, a short ten-minute walk from the station led us to the depot, and after going through the formalities of showing our permit, we were given free access to the shed and yard. The six-lane depot at Dundee had been converted from the former Caledonian steam shed and that day there were ten locos on shed, including three Inverness-based class 26s. Two stored 06 shunters were also noted having come to the end of their working life, most of which would have been spent shunting at the local docks.

Two withdrawn 06 shunters: 06005 and 06006 on Dundee depot during our Freedom of Scotland trip. Taken on the 30th of January 1982.

After the depot, we returned to the station on what was now a cloudy and cold morning; any earlier sunshine had most definitely disappeared, and there was (disappointingly) no sign of snow. From Dundee, we boarded the next service up the coast to Aberdeen, on which we passed a class 40 at Montrose. At Aberdeen, we made the twenty-minute walk to the shed at Ferryhill, where a modest twelve locos were stabled.

The shed itself dated back to the 1860s having been redeveloped by the Caledonian Railway, and later converted over to full diesel use in 1966. Its closure came on Boxing Day in 1987 and much of the site has now been cleared apart from the turntable and the original two-lane Caledonian building which is now in the hands of the Ferryhill Railway Heritage Trust.

The miserable weather we had left in Dundee gave way to bright sunshine in Aberdeen - the weather was definitely changeable and if it had been June, this would most definitely be T-shirt weather, but in January, it was freezing. Not helped by the biting wind which was coming in off the North Sea.

The other part of our great Scotland plan included watching a top drawer Scottish football match, and fortunately for us, Celtic were playing Aberdeen at Pittodrie. There was already some controversy about the game as the Celtic Supporters Association were boycotting the game due to Aberdeen's small ticket allocation for the visiting Celtic fans, so we were unsure how many Celtic (away) supporters would be in attendance. When we arrived, it became immediately apparent that the boycott had been largely ignored, and there were hundreds of Celtic fans queuing to get in at the away end.

Once inside, we became enveloped in the buzzing atmosphere, finding it easy to side with the Celtic fans. Aberdeen scored first, but Celtic came back strongly to win 3-1 in front of 19,000 supporters. Paul McStay scored for Celtic in this, his debut game. He would go on to become a legendary midfielder who would play over 500 games for Celtic and 76 for Scotland.

Others playing that day went on to become famous managers, including Gordon Strachan, Alex McGleish and even Mark McGee, who later ended up managing Bristol Rovers. Didn't see that one coming (it was a disaster).

The game was very feisty as the top two Scottish clubs fought it out and the singing from the Celtic fans was continuous and passionate, although I was a little put off by so many of them choosing to relieve themselves on the terrace as opposed to trekking to the toilets.

On our journey back to Glasgow, we were thus in the company of boisterous Celtic fans who were full of it after their win. Tonight was due to be our first night in the Bed and Breakfast on Sauchiehall Street, and not a moment too soon. We were both exhausted after the previous night's snatched rest, and I, for one, was looking forward to sleeping in a proper bed.

It's worth mentioning here that industrial disputes involving the railways were not uncommon in the 1980s, and at the time of our trip to Scotland, there was an ongoing dispute which meant no trains were running on Sunday the 31st of January. As we had known about this in advance we had planned it into our schedule and without access to any rail services would instead visit the sheds in the Glasgow area, making use of buses and taxis if need be.

The great thing about a strike was that with no trains running, the sheds would be fuller than usual and hopefully, with nothing moving, we figured it should be easier to get around. After a good night's rest, we filled up on a hearty 'full English' breakfast before heading out into a sunny but cold morning. We had plotted our route in advance, so aided with a Glasgow 'A to Z', we walked confidently in a westerly direction away from the city centre towards Hyndland depot.

During this time, some parts of the Glasgow townscape were transitioning from old-style tenements into more modern buildings. There was a lot of regeneration going on in the city as it attempted to discard its reputation for having some of the poorest areas of deprivation in the country.

Hyndland depot, which opened in November 1960, was located at the end of a short branch off the Helensburgh line and consisted of a three-lane structure capable of holding two three-car sets in each lane. Its primary role was maintaining the class 303 and 314 electric multiple units (EMUs), which plied the North Clyde suburban lines. Our walk to Hyndland took a little longer than anticipated, so there was much relief when we eventually found the depot entrance on Hyndland Road. As suspected, nothing was moving as a result of the strike, and only a few workers were located in the office close to the depot entrance. They were a cheery bunch, and when we mentioned that we were from Bristol, all they wanted to talk about were the race riots which had taken place a couple of years previously.

The race riots to which they were referring happened in Bristol as a result of a disenfranchised Afro-Caribbean community living in the St Pauls area who were protesting against a police raid on a cafe in the district. It seemed strange to me that the city I loved was now seen in the context of race riots, instigated as a reaction to government policy and the perceived poor policing of minority communities. These disturbances went on to trigger further riots in the Toxteth district of Liverpool and other inner-city areas and became a visual sign of the difficult times we were living in then.

Back at Hyndland, it didn't take long for the subject of the riots to be exhausted, and we were eventually given permission to go around the shed where we noted down fifteen units. Hyndland depot closed in 1986 when British Rail began reorganising its EMU maintenance arrangements in Scotland; the maintenance of these units now being focused on Yoker, a couple of miles away.

A couple of years after closing, the building was demolished, and today, the site has been redeveloped.

A general view on the 31st of January 1982 of Hyndland EMU depot at the start of our Sunday trek across Glasgow. The depot has since been demolished.

From Hyndland, it was a long trek back across the city to Eastfield depot. Our permit here was actually for the next day, but we hoped to be able to get in a day early as previous visits to the depot, without a permit, had always been trouble-free. The walk took forever before we eventually found ourselves on Hawthorn Street, where we followed the drive up to Eastfield shed. The offices were located at the south end of the shed whilst the bulk of locos were stabled at the northern end. I had never previously experienced any problems accessing Eastfield, although it looked as if good views were possible from the surrounding area if we were unlucky today. The foreman, though, was indeed friendly enough, and we went around noting down sixty-eight loco numbers.

27033 on Eastfield depot on Sunday the 31st of January 1982, taken during our Freedom of Scotland trip. A national rail strike meant that most sheds were fuller than usual.

A more general shot of Eastfield depot, Glasgow, on the 31st of January 1982.

Leaving Eastfield shed, we found our way onto Springburn Road, where we headed down to St Rollox Works. There was a security gate and entrance at the side of the works on Charles Street, and it was here we tried our luck gaining access. Both Brian and I knew it could be quite relaxed here, but even so, we were surprised to be allowed to wander unaccompanied around the whole site.

It was very quiet in the workshops, with everything just left where it was from the previous Friday. The twenty-six locos in the complex included classes 08, 20, 25, 26 and 27, a good traditional Scottish mix that reflected a time when locos were allocated to specific workshops for heavy repair. 47s would go to Crewe, for example, 37s to Doncaster and 45s to Derby. Some of the 26s and 27s we saw had already been withdrawn, and 27009 and 27044 were noted languishing in the scrap yard, looking a million miles away from the sunny days when they traversed the West Highland line skirting lochs on the way to Oban and Mallaig.

After visiting Eastfield, we managed to get around St. Rollox works. 20217 was stabled by the test area on the 31st of January 1982.

Another shot at St. Rollox works on the 31st of January 1982.

From St Rollox, we returned to Springburn Road, where, after waiting for a bus that never came, we hailed a cab to take us to Polmadie depot, which was next on our list. As we were walking down the incline into the depot, a car coming out of the shed stopped abruptly in front of us and the driver leaned out of the window. He then proceeded to tell us in no uncertain terms to 'get off his depot'! We were more than a little shocked to encounter such rudeness, especially considering the relaxed atmosphere we had experienced everywhere else. After he had driven off, we decided to double-check with the shed foreman just in case and found him to be much more accommodating, simply telling us to let him know when we had finished.

Polmadie depot was a forlorn and pale shadow of the former steam shed which had previously stood on the site. Originally constructed by the Caledonian Railway, it had comprised of a large running shed and a smaller repair shop however in the mid-1970s, the running shed had been demolished, leaving only a few stabling sidings and the repair shop, which is what we saw on our visit. There were about thirty engines on shed including withdrawn 06008, which was formerly D2437 - one of the earlier shunter designs built by Barclays in the late 1950s. We also noted a couple of withdrawn 25s and 27s looking as if they were there to keep the elderly shunter company.

We didn't encounter the angry railway man again, so, after thanking the foreman, we made our way off the shed and headed for our next stop, Motherwell depot. This time, we were a bit luckier with the buses and found one that dropped us off at Motherwell station which was only a ten-minute walk from the depot. This was another old, converted steam shed dating back to 1866 – a further legacy from the Caledonian Railway, which was stone built with arches above pairs of tracks leading into the shed. In 1982, it was still going strong, providing shunters to Mossend Yard (a couple of miles north of the depot) and to the mighty BSC steelworks at Ravenscraig. Both locations guaranteed steady work for Motherwell depot, and not surprisingly, at weekends, many locos could be found there. Our visit revealed forty-one locos, mostly consisting of 08s, 20s and 37s, although a Longsight-based 40 broke things up a bit.

Following the closure of the BSC steelworks in 1999 and a general decline in industrial activity, Motherwell shed became a lot quieter and eventually closed in 2007. Its fate was not helped by a serious fire, but surprisingly, it was reopened a few years later and is now used by Direct Rail Services as a support depot for some of their operations.

Time for another bus journey, this time from Motherwell to Hamilton, just a couple of miles away. Hamilton West shed was then a five-minute walk from the station of the same name. This was essentially a DMU depot with an out-stationed shunter from Motherwell and was another converted steam shed which was over a hundred years old. Increased electrification of the Glasgow suburban services meant its life, too, was due to be limited.

The depot closed later that year and its stock was transferred to Eastfield and Ayr. Like so many old steam sheds, Hamilton had bags of character; it was poorly lit with old stone flooring, a real nod to its origins. We recorded thirty-four DMUs on our visit, the bulk of which were Derby Works-built class 107 units with a few class 116 examples.

My favourite spot of the day, though, was hands down one of the Scottish Region class 122 single unit 'bubble cars', Sc55015. I liked these units because they were unusual and similar in appearance to the Class 121 single-car units that ran through Sea Mills (my local station), on the Bristol to Severn Beach line.

As it was January, daylight hours were short, meaning that by the time we left Hamilton, it was getting dark. We had completed our planned circuit, though, so took a bus back into central Glasgow, where we returned to our bed and breakfast before heading back out onto Sauchiehall Street for a few drinks. We still had some time to kill before bed and ended up watching *'The Exorcist'* in a fairly down-at-heel Glasgow cinema. Returning to the B&B through the cold, dark and now wet city centre streets after the film, caused us to feel a little uneasy - no doubt assisted by the sights we had watched of a girl with a swivelling head and God-fearing priests!

Monday 1st February 1982 dawned bright and breezy but still cold. Today was planned to be more relaxed in contrast to yesterday's shed bash tramping around Glasgow. We intended to take a service up to Inverness, head from there back to Glasgow, then go across to Edinburgh before catching a local service to Dunfermline, then back to Edinburgh, where we would pick up a late-night service down to Carlisle. Then, it would be a very early morning trip from Carlisle to Ayr. Writing this now, it doesn't sound like a particularly relaxed day!

Leaving our lovely B&B, we popped in to see what was on the blocks at Glasgow Central before heading over to Queen Street. Class 47, 47564 provided the power on the Inverness service as we entered the Queen Street tunnels and started the climb up Cowlairs bank. Leaving the urban sprawl of Glasgow behind, the train then took us up through Stirling and Perth and on into the Highlands - at last we were seeing some snow! Lots of it, particularly as we passed through Aviemore.

27017 standing at Perth on the 1st of February 1982.

Arriving at Inverness around lunchtime, we exited at the end of the platform into the depot, permits in hand, ready to have a good look around. Inverness shed, also known as Lochgorm Works, was originally the main workshop for the Inverness and Nairn Railway (and later the Highland Railway) and dated back to the 1860s. It was, and is, an interesting and attractive listed building made of stone, which still retains its original window frames. Unlike yesterday, where our focus had been on the quantity of engines seen, today was all about the quality, and in particular, the Birmingham Railway and Carriage Works constructed class 26s, which worked the far north of Scotland lines. We were rewarded with seven of these on shed, all of them cops, along with three 27s and two 47s.

Knowing that services from Inverness to Glasgow were not that frequent, Brian and I left the shed and headed out into the town, walking through Queensgate and onto Bank Street, where we sat on some seats overlooking the swirling waters of the River Ness. We had picked up some local meat and potato pies for lunch and happily worked our way through these before returning to the station. We still had an hour or so before our train back to Glasgow so I picked up the current edition of the 'Railway Magazine' from the station newsagent. In those days, I would get this magazine (along with 'Modern Railways') religiously every month, and I still have several years' worth stored in the attic at home. They have faithfully followed me around every time I've moved houses – I just can't quite bring myself to part with them.

26046 poking out of the shed at Inverness depot. The class 26s were the reason many enthusiasts trekked up to Inverness. Taken on the 1st of February 1982.

47564 on Inverness depot on the 1st of February 1982.

Awaiting departure, we sat in our carriage drinking tea and eating a couple of individually wrapped fruit cake slices which were a staple of every Travellers Fare buffet across the network.

Once we got to Glasgow we nipped on a service across to Edinburgh. It was now around 6pm which meant that Waverley was buzzing with commuter traffic and the night had again drawn in.

From Waverley, we had planned to take a local service out to Cowdenbeath, which would give us a view of Dunfermline Townhill yard and depot. Although it would be dark and the shed was set back from the main line, we thought it might be worthwhile, nonetheless. It would also serve to kill a few hours before we needed to board the late-night service from Edinburgh to Carlisle.

The trip to Dunfermline Townhill yard was indeed worthwhile. Though we only saw three class 20s, one was a cop for me which was an unexpected bonus. A couple of years earlier I had visited Townhill for the first time as a bolt-on to a Bristol to Edinburgh excursion. On arrival at Waverley one of the local Bristol spotters (Stan, who was a bit of a legend) had arranged for a coach to meet the train and follow a mini tour taking in Haymarket, Grangemouth, Millerhill, Townhill and Thornton Junction. The price was very reasonable, and whilst it was never clear how many permits were arranged for these trips, Stan always seemed to get the right result when dealing with officials. This trip had taken place in March, which, in a surprising contrast to the trip Brian and I were currently on, had been quite snowy, and the coach arrived at a locked up and apparently deserted Townhill shed. Stan, ever the hero, managed to attract someone's attention, though, and after a little gentle persuasion, we were given access.

Townhill shed had provided power primarily for the coal traffic which served the local collieries and Kincardine power station. It opened initially as a wagon works, but in the late 1960s, it was partly converted to a diesel maintenance facility, meaning the shed at Dunfermline Upper, closed. Townhill was nothing spectacular to look at, however, and there was no rich railway architecture to shout about.

Some thirty-five years after Brian and I visited, my job took me back to Dunfermline, where I ventured further than the engine shed and found the town to be very interesting. It benefits greatly from being the birthplace of the Scottish philanthropist Andrew Carnegie, who made his money from steel manufacturing in America. As part of his philanthropic work he funded the Dunfermline Carnegie Library, which was the first of many free libraries he built around the world, and he also funded the marvellous Pittencrieff Park– both crucial pieces of history which had passed me by on my earlier visits.

Dunfermline was also the hometown of the Skids, one of my favourite punk bands of the 1980s, so there were many reasons for me to enjoy it.

As with the rest of the country, freight traffic in Dunfermline (and Scotland as a whole) declined through the 1980s with Townhill depot, like the collieries and power stations it served, eventually closing before being demolished in 1989. Brian and I had been fortunate to see the depot one last time, even if only from a passing train.

Our trip back to Edinburgh after Townhill was uneventful, and we retired to the buffet at Waverley for more reheated sweaty pasties and tea before going on to Carlisle behind 47464. Here, we had to endure quite a wait before the next morning's first train up to Stranraer. It was freezing cold, and the lack of sleep was starting to take its toll. I tried unsuccessfully to nod off in the waiting room, but the benches were hard, and the lights too bright. Carlisle Citadel is undoubtedly an attractive and distinctive station, but its charms were unfortunately lost on me that long and cold night.

Eventually, the morning arrived, and we were able to board the Stranraer train, where sleep came quickly in the luxury of a heated compartment. Sadly, this meant that Kingmoor depot and yards were missed, but I did manage to open my eyes long enough to catch the shunter at Dumfries. Sleep then took over again, and the next thing I knew, we were heading into Stranraer.

The railway line here curves past the former Stranraer Town station, which was now a freight terminal, before ending at Stranraer Harbour station. We'd heard there was an 08 somewhere, but Brian and I didn't see it, although we did see a class 27 lurking in the yard. Continuing onto Ayr, both of us still half asleep, we jotted down the numbers of the class 126 Swindon built Inter-City units stabled in the station. These DMUs, which rattled between Ayr and Glasgow, were quite distinctive having different ends, one with a gangway and one without.

The closest station to Ayr Depot was Newton on Ayr, so this is where we left our warm and cosy compartment behind. Adjacent to the station was the large Falkland Yard, which was a focal point for coal traffic in the area, and this particular morning, we noted a Gateshead class 40 stabled. It was about a ten-minute walk from the station to the shed.

Framed by the floodlights of Ayr United's ground, Longsight based 25080 on Ayr depot. The picture doesn't convey just how cold it was! Taken on the 2nd of February 1982.

As it was still quite early, the streets were only just starting to come to life with people beginning their daily routines. Nobody paid much attention to two bleary-eyed train spotters walking down the road who were, by now, desperately trying to get warm on what was developing into a blisteringly cold Ayrshire morning. We'd known that the biting wind coming in off of the Firth of Clyde would take no prisoners and the tiredness combined with the icy temperatures were starting to have an effect. It was comforting to know this was our last day, after which we would be returning to Bristol and hopefully milder climes.

On arrival at Ayr depot we showed our permit to the foreman and began our visit. As it was a weekday, the shed was quiet with only a handful of engines; however, 27008 was present, which was, rather satisfyingly, my last class 27. Happy, (but still cold!) we returned to Newton on Ayr station to catch the next service to Glasgow, which passed by Shields Road Depot with its lines of class 303 electric units.

From Glasgow Central, it was a short walk over to Queens Street, where we boarded the first available Edinburgh train. We were going to alight from this train at Haymarket station, (the last stop before Waverley), as we had a permit for the depot of the same name. This was about a ten-minute walk from the station and the depot comprised of an odd collection of buildings, with some dating back to steam days. Haymarket was a key Scottish depot that provided locos for the London services, which meant it had the mighty Deltics as part of its allocation. Unfortunately, there were no Deltics to be seen today, though two Eastern region class 31s were keeping company with the 26s and 47s.

Whilst we knew we wouldn't find vast numbers of locos here – unlike at Eastfield – we knew we would see some quality engines. We didn't have enough time to trek out to Millerhill yard after our visit to Haymarket, so we spent the rest of the morning at Waverley before returning to Bristol on a homeward journey that took a good six hours and provided some fifty-odd further engines.

I was tired, worn out, hungry and cold and couldn't wait for my bed and a good, long, uninterrupted sleep.

Edinburgh Haymarket was the last shed we visited before heading back to Bristol. 31106 was stabled in the yard on February the 2nd 1982.

Towards The End Of The Century

One of the things I've learned about myself is that I am not very good at giving up on lost causes, or do I just have a steely determination to try and get what I want? In October 1983, I met Jackie and thought I had found true love. We got engaged and spent a couple of weeks the following summer riding my Honda CX500 (I had upgraded from the 250cc Superdream) down the west coast of France. One evening, we were in the charming Southwest Brittany town of Piriac-sur-Mer, sat outside a bar watching the sun set in a glorious display of ferocious reds and oranges, and, unbelievably, in the background, 10 CC's *'I'm not in Love'* drifted across the evening calm from a nearby apartment. Life was idyllic.

Six months later, it wasn't. Things were turning sour, and the spark had gone. During an Easter break, we reached the end of the road, and in the drizzly gloom of a North Cornish coastal town, we decided to call it a day. I was heartbroken. The split had been brewing for a few months, yet it happened during one of the few occasions we had been able to get away on our own. There I was reading the 'Railway Magazine', Jackie was reading a fashion magazine, and we weren't talking. This spoke volumes about our relationship. We vowed to stay friends, which actually made things worse. I wanted her back and was convinced I could win her over. We continued to see each other a couple of times a week over the next year – just as friends.

I was playing the long game, and my plan was to go away with Jackie for a fortnight's inter-railing around Europe. It was something we'd talked about when things had been better between us, and now I saw it as my last chance to find out if we could make it work.

On 22nd June 1985, Jackie and I set off with our backpacks, clutching two inter-rail tickets and our passports. The first part of our journey took us from Bristol to Dover via Victoria (73123 noted on shed at Stewarts Lane depot as we travelled through south London), and then onto the ferry (as foot passengers) over to France.

When we arrived in France, our first stop was Calais, where we stayed in a hotel for one night - bizarrely called the Hotel Bristol. The next morning, we set off for Paris and arrived at the Gare du Nord at mid-day before making our way over to the Gare d'Austerlitz for an overnight train to Madrid. Strangely enough the relationship, or lack of it, didn't seem such a big deal anymore and we ambled happily through Europe as two good friends.

The journey to Madrid was long and we drifted in and out of sleep in the comfy compartment. When we eventually reached Madrid's Chamartín station, it was a bright, sunny and hot June morning and, in the yard just outside of the station were a couple of former British Rail Class 14 'Teddy Bears'. Bonus! These locos had been exported to Spain at the end of their career on the Western Region, leaving our shores before I had become interested in railways. It seemed weird to see former Bristol Bath Road locos in Spain.

Leaving the station we made our way to the hotel. En route, a scruffy looking woman approached me asking for money which was a first for me and I was a little shocked. At the time, street begging was rare in England though some forty years later it would become a common sight on Bristol's streets. So much for progress.

Much of the next day was spent in Madrid but I don't think we made the best use of our time being unprepared both for the sights and the unrelenting heat.

SNCF Class BB 9288 at Paris Gare d'Austerlitz on the Inter Rail trip to Morocco. Taken on the 23rd of June 1985.

After Madrid it was onto Gibraltar via an overnight service, the train heading south through a hot balmy night where thunder rumbled, and an electrical storm crackled, lighting up the sky. At some point during the night our train stopped in the middle of nowhere. Jackie and I were bemused to watch people getting off the train and sitting down by the side of the track. Some were having a smoke, others stretching their legs. We had no idea what was going on but everyone seemed chilled and relaxed and there didn't appear to be any cause for panic. About an hour later, the engine sounded its horn and everyone got back on the train. I couldn't imagine this ever happening in Britain.

The following morning, we arrived at Algeciras at the Southern end of Spain, near to the Strait of Gibraltar. From the station we caught a bus to La Linea, which was the Spanish border post for crossing over to Gibraltar, then we simply walked out of Spain and into Gibraltar, which is a British Overseas Territory (and no longer in the EU following Brexit). The main road into Gibraltar was an eye opener though. It involved crossing the main airport runway – but only when no planes were due!

Our hotel was situated close to the centre of the island, and it was here that we stayed for a few days. Gibraltar was much like I imagined England to be in the 1960s; it had a lot of charm and faded grandeur with Mediterranean overtones. On one of the days, we took a trip to Morocco, where we were 'befriended' by a local guide at the ferry terminal in Tangiers. The guide helpfully whisked us around the old Kasbah, the Berber market and then into an old decrepit taxi for an interesting (and scary!) drive along roughly surfaced roads to the small coastal town of Asilah. The bruised and battered Mercedes taxi created a huge dust storm as it bounced along the poor-quality carriageway. Whilst Asilah was very picturesque, it was still incredibly hot and it was a relief to get back onto the ferry and feel the cool Mediterranean breeze again.

After Gibraltar, we took the train back up through Spain to Madrid. This was via another overnight service, but this time, we shared it with a trainload of Spanish soldiers heading back home from the Spanish territory of Ceuta on the North African coast. The toilet in our carriage had decided not to work, and with so many passengers, it didn't take long for the stench to become unpleasant.

From Spain, we headed back to France, where we spent a couple of days at Arcachon, which was reached via a local service from Bordeaux. Arcachon was (and probably still is) a trendy Atlantic coastal resort with a laid-back feel; however, the friendship/companionship thing was beginning to wear thin, and both mine and Jackie's moods began to deteriorate. The pretence of just being friends was getting harder to maintain, which meant that we both ended up being sulky. Increasingly so, the longer the trip continued. In the end, returning to England early became an attractive option and so we retraced our steps via Paris, Calais, Dover and eventually Victoria, arriving back in England two days earlier than planned. As we walked across central London to Paddington, the heavens opened, drenching us in a monsoon-like downpour. It was Wimbledon fortnight though, so no great surprise, and a 17- year-old named Boris Becker was about to become the youngest ever player to win the men's singles.

Once back at Temple Meads, we caught a local service to Severn Beach, where I got off at Sea Mills and Jackie at the next stop, Shirehampton, where she lived with her mother. We parted, still just about friends, but that was it; the long game had not worked.

It's not in my nature to be downbeat for too long, though, and in some ways, the holiday made things clearer. It was time to move on. The next day, 6th July 1985, was Cardiff Canton open day, one of many events that summer which marked the 150th anniversary of the opening of GWR's first line out of Paddington. The open day was shared between Cardiff Canton and Cathays, a former GWR steam shed which was now used for wagon repairs, although, I must be honest, Cathays didn't really interest me.

It was a warm sunny day and Canton was incredibly busy with queues forming to climb into the cabs of the visiting locos and the merchandise stalls were doing brisk business. A pair of Tinsley based class 20s, a Toton 58, and a southern 73 were all noteworthy visitors that had been brought in for the special day along with some preserved diesels; notably a Class 40, 40122, which was the first class 40 built by the English Electric Company in 1958 (originally numbered D200 but subsequently renumbered to 40122 later in life) and class 44, D4 Great Gable, one of the original ten of the peak class named after mountains. I was impressed. The organising committee had done a great job and no doubt some local charity benefited handsomely. Plus, I had found the perfect balm for my broken heart.

As part of the GWR 150 Celebrations in 1985, there were several open days across the region. On the 6th of July, Cardiff Canton Open Day took place. Here is Toton based 58003 on display as one of the many exhibits brought in for the event.

Moving on a couple of years from the 'defining' inter-rail trip around Europe, I met and married Liz, which resulted in a significant scaling back of activity on the railway front.

Southern Region visitor 73123 at Cardiff Canton Open Day on the 6th of July 1985. The loco is still operational with GBRF as 73963.

However, as I got into the swing of a stable married life, opportunities once again presented themselves to venture out, but things were different. The days of exhausting trips with the sole purpose of tracking down engines at far-flung outposts around the rail network were no longer so attractive. My life had changed significantly since then, and looking back, I think those trips had started to become more like endurance tests; how many nights could you manage with only snatched sleep here and there?

The rail scene itself was changing as well. In 1982, the regions upon which British Rail had based its operations (Western, Eastern, London Midland, Southern and Scottish) were abolished and replaced by 'business sectors' in a process known as *sectorisation* – the idea being that these 'businesses' would 'turn a profit'. Several passenger and freight sectors were formed, with each having its own brand, often resulting in specific liveries for their engines. Since the early 1970s nearly all British Rail locomotives had been painted in BR corporate 'blue', but that was to change as engines began to display the new liveries. I was twenty-six years of age at the time and the hobby which had occupied large chunks of my adolescent years was beginning to lose some of its appeal. I hadn't thrown the towel in as such, but other things were now jockeying for my time.

Despite all of these changes, one constant was Bristol Rovers Football Club, which still played a big part in my life. I began trying to visit new football grounds and would still add in a bit of railway activity for good measure. These visits also provided the opportunity to meet up again with Brian, who had moved away from Bristol. Which brings my story to the 21st November 1987, when Rovers were playing at Bury just north of Manchester and I was meeting up with Brian for a day out.

As was customary, it was an early start, and I took the 06.25 HST (not quite the 06.57 of Portsmouth FC fame) from Parkway up to Birmingham, where I was to meet Brian. Following the closure of Severn Tunnel yard, there was now increased freight traffic at Stoke Gifford yard (adjacent to Parkway station), which generated enough work to require a class 08 shunter from Bath Road depot to be based at the yard. After catching up with Brian at New Street, we set off in his ever-trusty Talbot Sunbeam car to Saltley and then negotiated Spaghetti Junction in order to visit a dull and overcast Bescot before undertaking further stops at Derby, Toton, Tinsley and Healey Mills.

56068 stabled on Saltley depot on the 4th of October 1986.

When we approached the driveway into Toton, instead of traversing the rail crossing into the depot, we turned right towards the little fishing lake situated at the Northwest side of the site. Here the path gradually fell away to rough ballast, which put the Sunbeam through its paces, but we were able to see a lot of what was on shed. At Tinsley - where there were noticeably fewer locos compared to previous visits – the highlight was the green liveried and preserved Peak D100 stabled down the side of the shed. 47207 also looked smart, displaying the new grey rail freight distribution sector livery (I still prefer blue, though!) plus there was a line of nine 08s in the depot yard. A worthwhile stop.

Against a leaden sky, we continued our journey across to West Yorkshire and onto Healey Mills. The downturn in the economy, particularly the steel and coal industries, was evidenced by another quiet yard and depot, yet, in keeping with tradition, we were still not allowed around the shed with the foreman again showing us a list of what he wasn't allowing us to see, I think he meant well, or did I see a glint of smugness in his eye?

In the end, it wasn't a complete waste of time. I copped one of my last class 37s and a 56 after a bit of a muddy trek across a field; both locos being tucked away in the yard. Lunch was simple: fish and chips from a nearby shop which only had three options on the menu: fish, fishcake and chips.

Bradford was our next stop, and here we picked up a couple of Brian's mates from Uni before continuing onto Bury and to the football match. We were full of hope in anticipation of a Rovers victory as we arrived at the footballing stronghold that was Gigg Lane. However, our hopes were misplaced. We got stuffed 4 - 1 in front of 2,356 poor souls braving a drizzly Lancashire afternoon.

We decided to drown our sorrows with a couple of pints in Manchester before I was dropped off at Piccadilly whilst Brian stayed on with his mates for the rest of the weekend.

The journey home was uneventful except for a football fight at Crewe station between Crystal Palace and Birmingham City hooligans. Oh, and copping a Scottish class 26 in the yard at Crewe Diesel Depot, which was a very welcome surprise.

Fast forward to March 1988, I had now been married for seven months and was living in a small Victorian terraced house in Brislington, a suburb of south Bristol. Liz was three months pregnant, and we knew that our lives were going to change significantly with the birth of our first child, so a weekend boys' break was planned that would take place before the 'big event'.

Class 86 electric, 86214 Sans Pareil at Manchester Piccadilly on the 3rd of July 1980. The loco was scrapped by Ron Hull at Rotherham in 2006.

I booked a couple of places on a STARS (Severnside Travel & Railway Society) trip that had secured permits for Derby Works and Toton for Saturday, 26th March which Brian and I were going to combine with watching Rovers at Doncaster on the previous Friday night. An overnight stop in Bradford was planned with Brian's mates.

Little did we know but this would be the last time Brian and I went on a serious rail trip together. Looking back, this trip marked the end of an era and drew a line in the sand under the previous ten or so years of bunking around sheds across the rail network.

The day started well, bright and dry, and I walked down to Temple Meads from the flat I was living in just off of Park Street in Bristol city centre, carrying my ever-faithful black Adidas sports bag which contained the bare minimum for an overnight stay. There was nothing much to see at Temple Meads other than a slightly out-of-place class 33. It was heading up a rake of empty carriages from the overnight sleeper train from Glasgow to Bristol, waiting for the green light to leave the station for the carriage sidings at Malago Vale to the south of Temple Meads.

The journey up to New Street frustratingly produced only two locos, which was also a sign of the times. Fifteen years ago, I would have seen many more on that stretch, including a couple of 37s at Bromsgrove waiting to provide a hand to help push heavy trains up the famous Lickey incline. I would have also possibly seen a 47 in the yard at Kings Norton on a car train, but today, there was nothing.

Birmingham New Street, to me, is one of the most uninspiring stations in the country. It is a concrete subterranean world devoid of any character, although perfectly suited to the endless stream of units and electric hauled services that grace its platforms. Amusingly, it has been called Mordor by railway enthusiasts, comparing it to the evil (fictional) realm of Sauron in J.R.R Tolkien's, *Lord of the Rings*.

From Birmingham New Street, I made my way to Duddeston, taking the first local service out to Walsall. Duddeston used to be an important station with a wagon repair workshop on the west side of the station and a carriage shed on the other- but now things are a good deal quieter. A few years before my visit in 1988, there was also an electric loco stabling point just to the north of the station, which usually produced a couple of 85s, but that had also gone.

Saltley shed was only a ten-minute walk away from Duddeston station down Duddeston Mill Road. There were thirteen locos on the depot that day, which represented six different classes - par for the course for Saltley. After twenty minutes or so, Brian pitched up in his still-faithful Talbot Sunbeam, and we headed north, calling in at Bescot, Shirebrook, Worksop and Doncaster, where we visited the station before going to see if anything was visible in the works yard.

33029 at Temple Meads on the morning of the 25th of March 1988.

The easiest way to see the works yard was to look through gaps in the railings outside its main entrance on Hexthorpe Road, and our efforts were rewarded with a 56, a couple of 31s and some 08s. We then continued down to the diesel depot but neither of us felt inclined either to ask for permission to visit or even to just bunk around - I'm not sure why. Ten years ago, we would have slipped past the offices at the front of the shed and made our way in through the back of the repair shop without even thinking about it. Perhaps as adults, we now lacked the courage of our youth, or maybe we just couldn't be bothered.

An hour later, Brian and I made our way into Belle Vue, the old home of Doncaster Rovers FC. One of the best things about having a passionate relationship with a football club is that feeling of expectation before any game. This expectation increases in intensity when you are watching your team away – perhaps due to the additional exertions you will have endured to get to the match. Tonight was no exception.

Away-game optimism is interesting. As a counterweight to this (often misplaced) optimism, I also regularly experience a strong sense of 'what the hell am I doing here?', which usually begins about twenty minutes into the game if things aren't going to plan on the pitch.

On this night in Doncaster, however, I can say with some pride that I was part of one the lowest League crowds that Rovers have played in front of. The number of spectators: 1,311. The result: 1 – 0 to us.

The final whistle was met with much relief from the small crowd on what had become a cold evening, and the ground quickly emptied. The majority of the crowd were disappointed with the result, but not us as we headed for a night out in Bradford with Brian's mates. The journey to Bradford didn't take long, and in no time at all we were in a city centre bar, enjoying a couple of beers before going to a nightclub. Flushed with Rovers' victory, we were in high spirits.

The club was a typical 1980s venue - lots of chrome to compliment the synthesised rock. A few too many drinks later, and the evening started to blur, so we left the club on a quest for food. The curry house we ended up in had come highly recommended, though it was very different from what I was used to in Bristol. No plush red upholstery piped sitar music, or soft lighting here. Instead, it had a linoleum floor, Formica tabletops and bright lights. To add to the ambience, the owner's family all sat eating at one end of the room, and when our food arrived, there was no cutlery provided, just a few chapatis to scoop it up. The food tasted fantastic and after eating our fill, we stumbled back to one of Brian's mate's houses and crashed out in the front room.

It is often said that we reap what we sow, and the following morning, which dawned grey and wet, was a perfect example of this. I awoke with a thumping head and nausea. The combination of six males sleeping in one room after a night of beer, curry and cigarettes created a fug which would have tested even the strongest stomach. I genuinely believe that if one of us had lit a match, the whole house would have gone up.

Despite our fragile condition, Brian and I still had our commitment to meet members of the STARS railway society outside of Derby Works in a few short hours; all we needed to do was make sure we got there in time. With the car windows wide open, we began our journey down the M1, stopping off at Tinsley before arriving at Derby Works.

Ten years ago, Tinsley had been a very busy yard but today we found things much quieter. This was due to several factors, not least the rapid decline of the coal industry as a result of Margaret Thatcher's drive to crush the National Union of Miners. Whilst I had no argument with the economics of this measure (fossil fuels had definitely had their time), I wasn't a big fan of the way the change was implemented using police as enforcers of political policy. Surely this is never a good idea. The combination of the eventual closure of almost every mine in the Yorkshire and Derbyshire coalfields, the destruction of once strong communities and the fact that the police were charged with maintaining law and order and protecting coal miners who wanted to continue going to work, left deep scars in the region which remain to this day.

I was saddened on this Tinsley visit to see the yards already in decline and many of the sidings overcome with weeds. Nevertheless, we parked up and started to climb the grass embankment at the side of the shed. Given its run-down state, we weren't optimistic that we would see the number of engines that we had come to expect at this location and we weren't wrong. The result was a single class 20, a collection of 08s, 31s, 37s, 45s, 47s and a solitary 56. We did see a few more unidentified locos lurking in the shed though.

A few years after this visit, the run-down of Tinsley would be complete. Sadly – though I was unaware at the time – this would be my last visit to what had once been a busy and iconic spotters' destination. The shed was closed and demolished ten years later and the yards reduced to a small dead-end siding serving nothing but a local steel plant.

Returning to the car that day, I was pleased to find my thumping head beginning to subside. Heading South on the M1, we reached Derby with a good hour to spare (before we were due to meet the STARS people), meaning there was enough time for some desperately needed coffee and a pasty – make or break time for the hangover. The pasty was a success, and feeling slightly better, we trudged out of the station buffet, making our way to the works, where we met up with the fresh-faced STARS enthusiasts who were clearly in better shape than us. After a short wait, an official guide emerged from the admin office and led us into the complex.

08492 under repair at Derby works, seen on the STARS visit on the 26th of March 1988. The 08 was the second from last shunter to be overhauled at Derby. The last one, 08856, was also seen and left the following month, marking the end of loco repairs at the works.

This works too was on its last legs. No longer would there be fifty-odd locos in various stages of repair. Today, we saw more stored and withdrawn locos in the scrap sidings between the works and the adjacent Etches Park depot than in the actual workshops. In fact, there were only three 08s in the shops, together with one HST power car, 43105. The grey rain clouds persisted overhead, which only added to the depressing state of the works, and we hoped fervently that a visit to Toton, a few miles down the A52, would help to lift the gloom.

Another photo from the visit to Derby works with more class 20s and shunters 08474 and 087223, all withdrawn. Taken on the 26th of March 1988.

Leaving our fellow enthusiasts behind (albeit temporarily), we made our way to Toton in Brian's car. Toton was still a major depot, and although the yards were also in decline, there was a fair amount of activity around the depot. Brian and I clocked fifty-five engines on shed, including eighteen class 20s and fourteen class 58s.

58050 Toton Traction Depot resting at Toton depot on the 26th of March 1988.

We were also able to see a line of rusting locos, mostly Peaks, residing in the yard to the south of the depot – many of which had passed through Bristol in earlier years heading up such iconic expresses as 'The Devonian' and the 'The Cornishman'. Also in this line-up was class 25 25080 which was one of the first of the class to transfer to Bath Road in January 1972 when the 'Hymeks' were being replaced. 25080 would go on to be moved around the country over the next few years before being eventually dismantled in Glasgow in 1993.

Finishing our visit to Toton, we bid farewell to the STARS party and decided we had time to catch another football match. A check of the newspaper led us to Villa Park where Villa were due to play Stoke City. Ironically, Villa's main striker that season was Garry Thompson, who would later, unimpressively, manage Rovers and take us down to the bottom division in the Football League for the first time in our history.

A long line of withdrawn Peaks and a couple of 25s waiting to be scrapped at Toton. Taken on the 26th of March 1988.

After finding somewhere to park, we entered the stadium and found ourselves standing on the mighty Holte End terrace. It wasn't as packed as when Villa had been in the old First Division, but it was still an impressive terrace full of partisan Villa fans. It was not a good day to be a Villa fan though, and after missing a hatful of chances they lost to a solitary Stoke goal. We left the ground and headed to Saltley for the second visit of the weekend, where we were able to notch up another seven class 58s.

And so, it ended. That weekend. It had been busy with a mix of trains, football and fun, but it wasn't just those two days that ended back then. So, too, had the boys' trips, which had become a staple part of mine and Brian's lives for such a long time. And now a digression from the railway theme, but there is a link.

At 04.14 on 15th November 1942, Kapitanleutnant Adolf Piening fired four torpedoes from his U-boat, U155, into the convoy he had been tracking since it had left Gibraltar, before diving down to a depth of 200 metres. It was a dark, calm night in the Atlantic, and whilst he did not know it at the time, his torpedoes scored a direct hit on HMS Avenger, an escort aircraft carrier. The missiles hit her munitions store and blew out the centre of the ship, lighting up the night sky in a blaze of flames and explosions. The bow and stern sections rose in the air and sank within two minutes, and that was the end of Hubert Carter, my Grandad, and five hundred and thirteen other poor souls who lost their lives that night.

HMS Avenger had only been in active service since the previous September, having been hastily deployed as part of Operation Torch to assist with the British landings in North Africa. Her first duty had been to escort a convoy from Scapa Flow to North Russia (PQ18), and my gran and dad were due to meet Hubert when the ship returned to the UK after completing the convoy escort. That never happened as the carrier's planned stop was cancelled in order to take part in the North African operation.

Piening was one of the most successful U-boat commanders of WWII, sinking a total of twenty-seven ships, with HMS Avenger being one of his greatest achievements - though he didn't find out what happened to the torpedoes he had discharged until 1948, three years after the end of the war. U155 is now also on the ocean floor, scuttled at the end of the war off the coast of Northern Ireland with many other U-boats. Like so many other war widows, my dad's mum only found out the fate of her husband after receiving a telegram with the statement 'missing presumed killed'.

Fast forward fifty years, and on Sunday 22nd November 1992, Dad and I are standing at the Cenotaph on Whitehall with a small group of people we didn't know but to whom we were connected. It was a special commemorative service to mark the fifty-year anniversary of the sinking of HMS Avenger and an opportunity to pay our respects to those who had lost their lives that day. Befitting of the occasion it was a damp, grey, drizzly day. The traffic was briefly paused by a police officer to allow for the short service to take place, and then afterwards, we retired out of the rain to a pub on Whitehall for a drink and a chance to talk to some of the others who had attended the service. The rest of the day was then free.

It had been a few years since I'd been to London with Dad and we reflected on how our father/son relationship had changed over time. Inevitably, following my marriage and the birth of George and Matt, my life became a lot busier, which I guess is the natural way of things. I wondered whether Dad had missed the trips that we used to do together - he would never have said - but we had the gift of an afternoon free with no clear plans. So, I suggested we take a trip down memory lane and visit a shed.

Half an hour later, we were on a local Southern Region stopper heading for good old Hither Green.

It was sixteen years earlier that Dad and I had first visited Hither Green, and to this day, it holds a record for me as being the only depot where I copped everything I saw. Three shunters, ten class 33s (including four 'slim Jims') and five class 73s and all in British Rail blue.

On the day of the HMS Avenger anniversary, as we left the train and walked down the platform to Hither Green shed, it struck me how little it had changed.

The shed buildings looked older and a bit more dilapidated, and weeds in the yard wrestled for space between the sleepers, but otherwise, the only discernible difference was the motive power and the liveries. As it was a Sunday, the depot was virtually deserted. There were no palisade fences, no security gates, just a good old-fashioned accessible shed.

By this time, Hither Green no longer had an allocation of locomotives and had become an outstation to nearby Stewarts Lane, which was the home shed for many locos of what was now the construction sector of Trainload Freight. So, we had a standard BR blue shunter, a Network South-East liveried class 73, a red Rail Express Systems class 47, grey Trainload Construction classes 56 and 60, and grey rail freight classes 47 and 33. But to be honest, apart from the newer class 60s, it was the same old engines, just repainted.

56051, 47296 and others stabled at Hither Green on the 22nd of November 1992. This was on the day of the memorial service for my Grandad (and others aboard the HMS Avenger) who died in WWII.

Seven years later, I would visit Hither Green again with Dad on a day trip to London, which we did annually for a few years to celebrate my birthday. This would end up being the last time I walked around a depot with him, more than twenty-six years after our first shed visit when we gingerly walked around the semi-derelict roundhouse at York. The man who had, on many fronts, been so instrumental in the path that my life had taken sadly passed away on Christmas Eve in 2020, just before his 89th birthday. He had suffered a short illness from which he never recovered. Like so many others at that time, the words COVID appeared on his death certificate; although it was a stubborn infection which was the primary cause of his passing - he had picked up COVID whilst in the hospital. Mum had passed away three years earlier, so now it was just my sister and I – orphans – with no one left to answer those old family questions.

Dad checking out the class 91s at Kings Cross on the 17th of April 1999. He was 67 years old. We didn't know it at the time, but this would be our last London rail trip together.

Rolling Into The Noughties

On 1 November 2000, there was quite an amazing railway accident in Bristol which, whilst thankfully resulting in no major life-changing injuries, dominated all the local news programmes. Due to heavy rain earlier in the week, the Chipping Sodbury tunnel on the main South Wales to London line, a few miles east of Bristol Parkway, had been flooded, which resulted in trains being diverted through Bath and Chippenham.

One such service was the 02.30 Avonmouth to Didcot coal service hauled by 60072. The loaded coal train was passing through Lawrence Hill station and in the process of negotiating North Somerset Junction before heading to Bath to re-join the main London line at Swindon, when it was hit in the rear by an empty mail train running from Parkway to Temple Meads. The mail train was top and tailed by 67002 and 67012, and the collision resulted in 67002 being forced on top of the last four wagons of the Avonmouth to Didcot freight before coming to rest just before the road bridge.

Following investigation, it transpired there had been a problem with the braking system on the mail train and it had allegedly passed through two red lights prior to the collision. The driver was extremely fortunate to suffer only a broken arm, cuts and bruises, but imagine the scene for both drivers.

The coal train driver would have felt a severe jolt and, realising that something had hit the back of his train, he then presumably left his loco and started to walk in the direction of Lawrence Hill station. As he traversed under the bridge, the first thing he would have seen would have been the class 67 perched on top of his coal train. Equally, the driver of the mail train must have feared the worst when he saw the tail light of the coal train coming towards him, only to find himself being launched into the air and then riding along on top of the hoppers.

67002 sitting on top of an MGR wagon following the mishap at Lawrence Hill, Bristol on the 1st of November 2000. An empty mail train ran into the back of a Didcot bound coal train. No one was seriously injured - which was some kind of miracle!

Surprisingly, the platforms at Lawrence Hill station remained open that morning and I was one of many taking photos. Looking at the shots now, it still amazes me how close I was able to get to the scene – and how lucky the driver of the mail train had been. The line was closed for a few days, and 67002 was lifted onto the back of a lorry to be carted off to Toton for major repairs.

I never had any great desire to visit the far-flung South Western extremes of the rail network; after all, anything moving in Cornwall would, at some point, turn up in Bristol. Even the Plymouth (Laira) based shunters have at some time mostly been allocated to Bath Road.

The old Penzance (Long Rock) shed, though, did pique my interest - I had seen photos of Warships and Westerns languishing outside the run-down steam shed - but I missed the opportunity to visit before it was demolished and replaced by the adjacent HST servicing depot. That really only left St Blazey, the unique roundhouse built in 1874 as the headquarters for the Cornwall Minerals Railway (which provided motive power for the local china clay traffic), as the one remaining place of interest.

In the early 1980s, I would go off on trips with a group of mates. Sometimes we would pop over to France or Holland on overnight coach trips - we even managed a cruise to Sweden, which, after spending twenty-four hours drifting past fog-shrouded oil rigs in the North Sea, left us with just three hours in Gothenburg before we had to return to the ship. But on a Sunday in August 1981, we found ourselves travelling a little closer to home on a special excursion which ran from stations on the Severn Beach line to Newquay, the surf capital of the southwest. The special was hauled both ways by Gateshead based 47410, and even though railway matters weren't on the agenda on these trips, I did catch a brief glimpse of St Blazey shed as the Gateshead 47 eased around the curve from Par station onto the Newquay branch. The depot was only partially visible, though, and the only loco I could note was local shunter 08113.

Twenty or so years later, I had the chance for a return visit. For a few years prior, our family holiday had been taken in Cornwall, or to be more precise, at Port Isaac on the North Cornish coast. Port Isaac is a gem. It's a former fishing village tucked away in a sheltered inlet, protected by high cliffs and a breakwater from the unpredictable North Atlantic swells.

The steep hills and narrow, twisty streets down by the harbour make visiting the place a bit of a challenge for a car driver - which probably helped to keep it off the map for the coach tour operators, at least until the popularity of the Doc Martin TV series took hold which resulted in a significant increase in the number of visitors in later years. As the kids were getting older, I went off on my own for a couple of hours and headed down to St Blazey. We were in the middle of one of the hottest summers on record, and even with the air conditioning on full blast, the car was still uncomfortable.

Just after leaving Bodmin, as the road dropped down towards St Austell, the English Channel appeared on the horizon in a shimmering haze. Some ten minutes later, I drove into St Blazey and easily found the depot. The roundhouse was no longer rail-connected and had a new lease of life as local industrial units. Any remaining railway activity was now focused on the former goods shed, which had been modified into a fuelling and servicing facility.

As an EWS outpost (English, Welsh and Scottish – major freight operator at the time), I was expecting to see a mix of class 66s that were used for china clay traffic and perhaps a 67, resting between parcel duties. The view from the former roundhouse of the fenced-off depot though was restricted, however, it was possible to see the other side of the depot from a footpath which runs near the old station down towards the shed. From here I was able to note one 08, three 66s, and a 67 rested up for the weekend. Long-time resident 47306 'The Sapper' - still in its two-tone grey rail freight livery – was also there having been withdrawn from service a couple of years earlier and stored at St Blazey ever since. A few years after this visit, 'The

A general view of St Blazey depot. In the distance can be seen stored 47305 which would eventually be brought back to life and end up on the Bodmin Railway.

Sapper' would enter the preservation world and take a short trip up to the Bodmin and Wenford Railway. It was great to see the depot but, after about half an hour, I didn't really have anything left to do, so I headed back to Port Isaac. Not long after this visit, the china clay and parcels traffic would significantly reduce.

Many people travel as part of their working day, and for some, it can be such an integral part of their jobs that it becomes a pain. However, I find a little bit of travelling here and there enjoyable, and it helps to break up the monotony of the daily grind. I had the chance to travel occasionally for work, and when I did, I would often try and tag something onto the day of a railway nature, perhaps a visit to a station or a shed. I guess in many ways, these visits were a bit of a trip down memory lane, and without wanting to become too melancholic, they were quite a good marker for how the railway world was changing.

This takes us to the 25th of July 2005 when I was heading up to Nottingham for a work meeting which was likely to be wrapped up by 4.00 pm. It should have been a glorious summer's day, but unfortunately, it was dull, overcast and cold. The journey from Bristol to Birmingham was uneventful; the Virgin Voyager was only half full. After leaving New Street, the unit negotiated the complex junctions to the east of the station before heading out towards the Freightliner terminal at Lawley Street and past the now-disused depot at Saltley. The Voyager pulled into Derby right on time, where I left it and boarded a two-car sprinter unit for the short run to Nottingham.

Making small talk, I discovered that the middle-aged couple sitting across the aisle from me were heading for a week's holiday in a caravan at Skegness and did not appear so happy at the prospect. The wife was stressing over what she had packed (or, more to the point, hadn't) while the husband was busy planning how they would spend the next seven days, none of which seemed to involve leaving the site or, more importantly, the on-site bar. No disrespect to Skegness but it didn't sound like a great holiday to me either, even if 'the air is so bracing'.

At Nottingham, I left them to their deliberations and headed out of the station for the short walk to the Council Offices where the meeting was taking place. By 3pm, it had finished, and I was faced with a dilemma. I knew that I ought to head home as I was trying to sort out the purchase of a car for Liz, which was to be a long-promised 40th birthday present. Having seen a suitable car the previous night, I was keen to finalise the deal. Also, Rovers were playing City in a pre-season friendly, and although I did not have a ticket, the kids were going and I would probably have been able to make it back in time for the last few minutes of the game, timing my arrival for when the gates were opened to let spectators leave. The dilemma, though, centred around paying a visit to Toton, something which I had been looking forward to in the run-up to this trip. I wanted to see how, what had once been the largest depot in Europe, had fared over the years.

Decision made (Toton won), I set about finding which bus service went closest to the depot. Ironically, it was the same bus company (Barton's) as when I had last visited the shed some twenty years earlier. I made my way to Victoria bus station in the city centre and boarded the Rainbow 4 service to Derby via Sandiacre. The outward journey was quick, and in no time, we were almost at Sandiacre. I sat near the front of the bus, keeping a lookout for the Midland Hotel in Sandiacre, which was the closest stop to the depot. Hotel spotted, I left the bus and found the path which led up to the famous bank that for years had provided spotters with such a great view of the shed and the marshalling yards. As I walked up the path, more of the railway landscape came into view. I didn't know what to expect but prepared myself for the worst.

Initially, I noticed that the large 'up' yard had all but disappeared under vegetation - which now resembled a small forest - and there was similar dereliction on the 'down' sidings. However, once up on the bank, the view over to Toton shed was as good as ever. No more class 20s and 56s to be seen, but plenty of 60s and 66s, even if many of the former were stored.

In fact, there were a lot of stored and rusting shells of engines littered around the site, so I spent the next hour enjoying noting down as many engines as I could, in a flashback to my youth. I then walked back down the lane off the bank and clambered up an embankment which led to the main A52 Nottingham to Derby dual carriageway. From this road, there was a bridge over the railway, which offered a great, if distant, view of the front of the shed and enabled a few more engines to be picked off and a few more photos to be taken.

Happy with my work, I indulged in some feelings of nostalgia, harking back to previous visits. I also had a sense, though, that the modern-day scene remained interesting. Returning to the bus stop, I headed back to Derby, where I took the next train to Bristol. I managed to get back in time to sort the car out but did miss the match, which wasn't a bad thing as we lost 1 - 0.

A general view of Toton depot taken from the A52 road bridge. At this time, class 60s dominated the scene. Picture taken on the 9th of October 2009.

Every September throughout the early 2000s, Bristol hosted a large second-hand railway collectors' sale at which almost anything connected with railways could be purchased. I had never been to one of these gatherings before and didn't know what to expect. For the past few years, I had been popping into second-hand bookshops looking out for old Ian Allan ABCs (booklets detailing all the engine and unit numbers), and every now and then, one would crop up. To me it didn't really matter whether the numbers were underlined by some previous owner or if it was unused, although I now know that the value of unused ABCs is significantly higher.

In many ways, I personally think the used ones are more interesting because they tell the story of someone's personal collection. It's even better if there is a name and address written in the front (which there often is), neatly scribed in the hope that should the book be left on a platform somewhere, a good Samaritan would take the trouble to return it to its owner. The prices of these books can vary; however, usually, after a bit of haggling, a sensible price can be agreed upon.

I had picked up the flyer for the Bristol event whilst checking out Bristol's Book Barn several months earlier. The Book Barn was a great place for people to take their unwanted books, and if they were subsequently purchased, the original owner would receive some commission in return. It occupied a huge warehouse and often had a good selection of railway books. Sadly it has now been demolished and replaced by flats.

The September 2005 collector's sale was taking place in a school about three miles from home, just by Parkway station, so I left the house with the kids still in bed. It was a hot morning and I had the car windows down, trying to get some fresh air in (also in part in an effort to rid it of the smell from the previous night's Indian takeaway which had spilt during the journey home). Though I didn't find the smell unpleasant, it was a little too rich at that time of day.

Enjoying some time to myself, I put on one of my newly purchased (remastered) Status Quo CDs ('On the Level', containing their one and only number one hit – Down, Down), and I let the thumping twelve-bar blues blast out over North Bristol.

Needing to get some cash, I drove past the school and popped into Bristol Parkway to use the cashpoint and, just as I was leaving, I heard a loud roaring engine noise coming from the station, much louder than anything you would associate with the modern railway scene. Looking back towards the station, whilst also trying to drive (I wonder if there will ever be an offence for watching trains whilst driving - I know that I have had many close shaves over the years), I saw D1015 Western Champion thunder past on a special service heading towards South Wales. The ochre livery looked fantastic in the bright sunshine. It must have been a good thirty years since I had seen a Western hauling a passenger train and almost as many years since I had seen an engine pulling that many coaches. It was a sight to behold.

Arriving at the fair, the first thing that struck me was the size of the event - there must have been at least thirty traders, all with their wares set out on trestle tables or displayed in portable bookcases. Everything you could imagine was on sale, including engine nameplates, posters, signs, number plates, timetables, books, tickets, uniforms, lamps, ticket punching things, model buses (not quite sure why they were there), photographs and a good selection of ABCs. This was big business, it seemed, and there was all sorts of wheeling and dealing going on.

I quickly found that if I hovered too long by a stand, I would be pounced upon by traders who could show second-hand car salesmen a thing or two, so I made sure to keep moving. As for the ABCs, a quick scout around revealed a wide selection up for grabs with a huge range in prices between the stands. One of the stalls specialised in the little books, with each one in mint condition and kept in a plastic bag. These were also the most expensive.

From a more reasonably priced trader, I picked up three excellent condition ABCs from the 70s and 80s for under £5 each, the prices here being noticeably lower than in Bristol's second-hand bookshops. I was feeling pleased with myself so grabbed a coffee from the makeshift café in order to examine my purchases.

The hall was busy, and it was noticeable, perhaps not surprisingly, that nearly every person there was male, white and generally over forty. As well as the enthusiasts like me and those with a general interest, there were some serious collectors in attendance. I overheard conversations referring to lifetime hunts for certain railway artefacts and stories were being recounted of former friends and characters from the railway fraternity.

Some people scoured through boxes of photos of steam engines looking for shots of engines – specific ones - whilst consulting handwritten lists of numbers on scrappy bits of paper or in little notebooks. I began to wonder if this is what happens when all the engines that you were interested in from your 'era' have gone. Do you start collecting photos of the engines because you can no longer see the real thing? Would I be doing this sometime in the future?

I left pondering and happy, with some new books to put on my shelf at home.

During a hot spell in June 2009, (and there weren't many of them), I went on an overnight visit to York for a two-day conference, conveniently held in the National Railway Museum. It had been nearly thirty-five years since my first visit to York so I was looking forward to revisiting the famous railway town.

The journey was fast and comfortable in another cross-country Voyager and I tried not to be too melancholic at the changed railway landscape flashing by my window. In particular the industrial backdrop north of Sheffield had all but disappeared having been replaced by a combination of business parks, housing and extensive landscaping. Who can possibly argue that is not an improvement? One of the most powerful memories of my first trip to York in 1973 was the bleak industrial landscape of coke plants, steel works and their associated slag heaps, particularly on the stretch between Sheffield and Rotherham. So, I had mixed emotions to no longer see it.

But enough reminiscing, the Voyager was approaching York on a glorious June day with the heat from the sun battling the train air conditioning. On arrival, I marvelled as ever at York station which remains an impressive tribute to our railway heritage even if the number of tracks passing under the iconic curved glass roof have significantly reduced.

I left the station and made my way to the hotel where I was staying, just by the banks of the River Ouse. It was lunchtime and the city centre was a mix of tourists and office workers taking a break in the warm sunshine and making good use of the riverside bars and cafés. The walk to the NRM (National Railway Museum) along Leaman Road was a well-trodden path for railway enthusiasts, and the gloomy subterranean tunnel underneath the line was in stark contrast to the brightness of the day.

The first day of the conference took place in one of the meeting rooms on the first floor and ended just as the museum was closing, so there was little time for exploring. At the end of day one, I walked out of the museum and turned right into Leaman Road as I wanted to go and have a look at the multiple-unit servicing depot, which had been constructed on the site of the loco stabling sidings at the top end of the former shed yard. The depot was of a prefabricated modular design and was shielded from the main line by a high fence. There were no units in sight; in fact, I didn't see anything or anyone at all.

The next morning, I arrived at the NRM early so I could have a walk around the museum before the conference started. Adjoining the main hall, on what must be the site of the former diesel depot (and may incorporate part of it), is the museum workshop, which can be accessed by a flight of stairs from the main hall. These lead to an elevated walkway which provides a view into the repair area. Unfortunately, as the museum had not yet opened to the public, the doorway leading to this area was closed; however, slipping back into habits from twenty-odd years ago, I opened the door and decided to have an 'unofficial' stroll around.

In the workshop was Deltic 55002, 'The Kings Own Yorkshire Light Infantry' receiving attention. It seemed ironic that this loco should be at its former home depot replicating what would have been a common scene in the 1980s.

The conference broke up earlier than the previous day, so, at midday, I had time for a less rushed look around both of the museum sites. Whilst the NRM has many railway artefacts from both the UK and around the world, for me, it was always about the diesels, and in one corner of the converted roundhouse stood a lovingly restored class 52 Western, shiny in its blue and yellow corporate livery of the 1970s. These engines had been at the very start of my railway journey, and it had been a good few years since they last operated under British Rail, so it was great to be able to get up close and personal with an example of what was my favourite class.

Sneaking around York Museum before opening time on the 22nd of June 2009, I found myself in the workshop where Deltic 55002 was receiving attention.

I broke the journey back to Bristol with a stop at Chesterfield to see how busy this bottleneck on the Midland mainline is today. In the seventies and eighties, when the coal industry was booming, and South Yorkshire was a major manufacturing powerhouse, Chesterfield station was a great place to while away an hour or so with guarantees of pairs of 20s, 56s, 58s and other classes passing through on a steady stream of rumbling freights.

Today, though, it was, as expected, much quieter, although just before my Voyager back to Bristol was due, 57006 pulled up at a red light with a train of scrap metal heading towards Toton. In a previous life, this reconstructed class 47 had held the numbers 47187 and D1837, and I had first seen it at Temple Meads in March 1974, thirty-five years earlier.

Now, as I watched, the driver closed the engine down and jumped from the cab to head in the direction of the buffet for refreshments. On returning, he picked up a loose brick from the line side and disappeared into the engine compartment. Shortly thereafter, a loud hammering could be heard as the brick was repeatedly smashed against some part of the engine before, eventually, the 57 burst into life. The driver returned to the cab, threw the brick out of the window, and eased the train forward. I cannot imagine that procedure was detailed in any driver's manual! I found myself a seat on the next Voyager back to Bristol and drifted off into a deep sleep, just managing to wake up as the unit eased into Bristol Parkway.

Western class 52 D1023 Western Fusilier in York National Railway Museum on the 22nd of June 2009.

A couple of months later, I fulfilled an ambition that had been niggling for quite a few years. I bought another motorbike. It was more than twenty years since I had last ridden, and motorcycle technology had advanced considerably over that period. The Suzuki SV650 I bought was much quicker than anything I had previously ridden and after a few weeks of getting used to it (not without some bottom-clenching moments I might add), I felt a bit more confident and decided to repeat a journey I used to make on many weekends in the early 1980s. I was going to visit Severn Tunnel Junction.

The depot had closed in 1987, and the yards had since been taken up. I had been there for the day of the depot's closure on 11th October 1987. The occasion was marked by a Bath Road Metropolitan Cammell DMU ceremoniously leaving the depot and setting off detonators in fine style, I have no idea where it went.

The only locos on depot on closure day were 31405 and 37691, along with the six-yard pilots, which were all lined up outside the small shed. The whole southern boundary of the site is now skirted by the re-routed M4 following the construction of the Second Severn Crossing bridge. I knew that repeating the journey would be more about nostalgia than anything else, but I was still looking forward to the ride. I chose a nice dry Sunday morning and headed off.

I had forgotten how noisy it is riding a bike at speed on a motorway and, before long, realised why many bikers today wear earplugs - not something I would have ever considered in my youth.

Leaving the M4 I rode down through Caldicot then turned left at Rogiet (the sign for Severn Tunnel Junction Station). The old local pub, The Rogiet Arms, was now, sadly, closed.

I rode up onto the bridge which had spanned the yards where I'd spent many happy hours in the 1970s, watching the 08s as they shunted wagons over the hump. All of this was long gone, although the diesel depot was still there, devoid of track and isolated. As I stood there, an HST hurtled under the bridge on its way to London, just as they used to do in the 1970s, representing a link from the present day back to the past.

My Suzuki SV650 represented a return to motorcycling after twenty-five years.

The following year, I swapped the Suzuki for a Kawasaki Z750 and then, in turn, replaced that with a Ducati 796 Monster. It had bags of character and made a lovely sound. I rode the bike down to Westbury a few times before deciding to go a bit further afield and ride down to Southampton. On the 1st of May 2015, on a lovely late Spring morning, I set off down the A36 to Southampton.

08836/760/822/785/589/848 lined up at Severn Tunnel Junction depot on the last day before closure. The depot closed on the 11th of October 1987, which represented the end of an era.

A sad looking disused Severn Tunnel Junction depot, but at least it's still standing. Taken on the 14th of November 2018.

After clearing the rush hour traffic in Bristol and Bath, the ride down through Limply Stoke following the Avon Valley to Warminster was perfect. I stopped for coffee and some breakfast just outside Warminster in a 'Little Chef' (iconic roadside café chain, not as commonplace today) before riding the delightfully scenic route to Salisbury through quintessential English villages such as Knook, Codford St Mary and Great Wishford. After bypassing Salisbury, I was soon on the outskirts of Southampton and could see the large dockside container cranes on the horizon.

The first stop was at Millbrook Freightliner terminal, where the road leading to one of the dock entrances passes over the running lines into Southampton. Here, you can get a good view of the small Freightliner depot where a few green-liveried 66s were stabled. By now it was quite hot and a bit uncomfortable under my biking gear, so I threaded my way across Southampton and arrived at Eastleigh half an hour later. It was a relief to go onto the station and take some of my riding kit off.

Eastleigh station is a busy place with a steady stream of Freightliner, aggregates and civil engineering trains passing through, and still attracts a hardcore of enthusiasts at the platform ends. I saw some seasoned spotters who looked settled in for the day, exchanging banter and insults in a way that only happens with people who have known each other for a long time. I don't think there was anyone under fifty years old.

My Ducati 796 Monster which I rode to visit the sheds at Westbury and Southampton.

Eastleigh is probably my favourite station; it is largely unchanged from when I went there in the 70s and retains bags of charm. After about an hour, I got a bit restless, so I jumped back on the bike and went down to the works and diesel depot, neither of which produced anything of interest. In fact, the view of the depot is now so restricted that it is hardly worth the effort of going down there. A high wall acts as a screen between Campbell Road and the stabling sidings, so there is literally no way to see anything other than a limited view across the front and rear of the shed. The ride back to Bristol was almost as enjoyable as the ride down, although there was a bit more traffic around, and the weather had turned cloudier and cooler.

A general view of the Freightliner depot at Southampton Docks with a few of the class 66s and 08 present. Taken on the 1st of September 2017.

66102 passing through Eastleigh on the 17th of April 2019 with a Northbound Freightliner service.

Colas liveried class 70s, 70808 and 70801 at Westbury Yard on the 9th of September 2016. Three other members of the class can be seen in the background.

A detour to Westbury (station) for a cup of tea and a leg stretch produced a couple of 59s on returning empty stone trains and a few Colas 70s stabled near the site of the old diesel depot. Again, this had long since been demolished.

Like Eastleigh, Westbury station has also retained its character, is still quite busy and is a good place to pass an hour or two. By early evening, I was back home, complete with aching knees and back, glad to be off the bike. I didn't know if it was just me or if age was finally catching up with me. Either that, or it's the fault of the Italian motorcycle designers for prioritising design over functionality.

Following my divorce from Liz, a few years later, I met and married Tracey who had also gone through a split from her husband. One of the great things about new relationships is that they open you to someone else's view of the world and their interests. Tracey likes cruises, which, I must be honest, never held much attraction for me; still, it's good to try new things, and after a couple of cruises to the Mediterranean and the Canary Islands, I decided I quite liked them, though I wasn't so keen on certain aspects like the formal dinners.

What I did really enjoy was waking up in a new place every day and being able to do our own thing after we had docked. Often we would choose not to go on the organised (and expensive) excursions, so for us, part of the fun was bartering a deal with a local taxi driver or just walking off the ship to explore on our own. In terms of cost, cruises can offer good value for money if you shop around.

On one occasion, in June 2015, we did a cruise to the Norwegian Fjords, which called at Bergen on the end of the rail line from Oslo. Although it was summer, Bergen was shrouded in mist and quite damp. Finding myself with a couple of hours free, I went in search of the railway station, which was a ten-minute walk from the town centre.

Bergen is an impressive railway station with an arched glass roof spanning several platforms. The station was very clean with an art deco feel and was painted in a muted grey. It was quiet for a Monday morning and there were only a few electric units stabled between services. Visible from the end of the station was a four-road engine shed which had a grey electric loco stabled outside, probably about five hundred yards away. I felt like I had gone back twenty years because suddenly, I didn't want to leave the station - not until I had got a closer look at the grey electric loco.

After about ten minutes, the loco started moving off the shed and came into the station. It was a class El 18 Adtranzk built (in 1997) electric loco no 18 2257 and looked quite sleek in its grey livery. I was pleased that I had seen it up close; more often than not, engines don't tend to appear when you want them to.

Also stabled in the station was a diesel-electric class Di 8 no 8.713, which was operated by Cargo Net. It reminded me of the old British Rail Clayton class 17 with its high central cab, although the yellow and red livery was more attractive than the drab green or corporate blue of its earlier British counterpart.

Twenty of these Norwegian engines were built, although ironically, ten were sold to GB Railfreight for use in the Teesside steelworks at Redcar.

Cargo Net operated diesel electric class Di no. 8713 in the station at Bergen, Norway on the 29th of June 2015.

A couple of days later, the cruise stopped at Flam on the Sognefjord. This is a hugely popular destination made famous by the Flam to Myrdal railway, which joins up with the Bergen to Oslo line. It offers spectacular views as the train threads its way through the valleys away from the Fjord. At least, so I've been told.

Unfortunately, we'd decided not to buy advance tickets for the train on the cruise and planned to get our own cheaper ones at Flam. This would have been fine, except the cruise line had block-booked nearly the whole train! So, there were no tickets available. (Note to self: don't try to save money in these situations). Still, I managed to salvage something and booked a couple of places on a local bus tour. Before the bus left I had a wander down to the station, where I saw four more grey class 18 electrics working the heavily laden tourist services. I didn't see any more locos on that particular trip, but it did plant a little seed, and now I keep an eye out for Norwegian railway articles in the railway press.

Norwegian State Railways class 18 locos 2245 and 2248 at Flam on the 1st of July 2015. We were on a Fjord cruise at the time.

Of all the train operators, Direct Rail Services (DRS) are probably the most enthusiast-friendly and every other year, they carry on the tradition of open days by alternating an event between their two main depots, Carlisle Kingmoor and Crewe Gresty Bridge. I decided to go to the 2016 open day, which was at Crewe.

Gresty Bridge is a relatively new depot, opened by Direct Rail Services in 2007 using buildings that were previously part of a wagon workshop. I decided to drive there to give me the option of visiting other locations in the area. I wanted to pop into Stourbridge Junction with the hope of seeing a stabled class 68 from the Chiltern line services. Back in the 1970s, I'd passed through Stourbridge Junction a few times on excursions from Bristol, and at the time, it had been one of the bigger freight yards in the Birmingham area. On a weekend back then there would usually be a few class 25s stabled between duties.

By the time of my visit in 2016, the freight yard had long gone, but some of the sidings had been converted for passenger stock stabling of both loco-hauled carriages and units. I found my way to the station car park and was rewarded with a view of the sidings where a grey liveried 68010 was resting at the end of a rake of carriages.

Leaving Stourbridge, I then got lost and, after several attempts, eventually found my way back onto the M5 just before the M6 junction. As I drove over the elevated section of the M6 at Bescot, I spotted a couple of 66s in the large and still quite busy yard before continuing on my journey to Crewe. The early morning mist had now given way to warm sunshine and I could tell it was going to be a fine summer's day. It was close to midday when I pulled off the M6 onto the A500 and headed for Crewe. The tall floodlights at Basford Hall, the large yard to the south of Crewe station, soon came into view.

A couple of years earlier I had trekked across boggy marshland to get a view of the locos in the yard, but a new road had now been laid closer to the yard, which gave access to a small residential development. With great foresight, the planners had placed a condition on the planning consent that a mound should be constructed between the houses and the yard, with a five-foot-high fence atop - presumably as an acoustic measure to reduce noise from the yard. Today, it made a great viewing gallery with about thirty enthusiasts craning for a view of the locos, many of which were tucked away between wagons in the yard.

I parked up and found that it was still possible to walk across the field to get a bit closer, though despite a few others having made the same trek, there was no clear path. The grass/scrub was waist-high in places, and with the addition of a few thistles, I soon realised it was not the best time to be wearing shorts! There were a lot of people milling around at the depot, although something slightly odd occurred at the entrance. I was enthusiastically handed an anti-nuclear weapons leaflet by an older gent who looked as if he was still entrenched in the 1960s. I have nothing against a bit of campaigning, but I couldn't really understand what the incentive was for targeting railway enthusiasts. Surely, it wasn't because DRS provided the motive power for the nuclear flask trains. Or was it?

For me, the two loco highlights of the open day were a mix of the old and the new. In the shed was an immaculate 68020, representing the newest class to appear on the network, and representing the old was a beautifully restored Deltic, D9009 Alycidon, a credit to the Deltic Preservation Society. Ironically, D9009 Alycidon was, back in the day, the last Deltic I needed to clear the class when I saw it in September 1977, resting on the buffers at Kings Cross.

Alongside the locos was a stall from Stadler, the manufacturer of the class 68. The enthusiastic company representatives were giving out freebies emblazoned with the Stadler logo which included postcards, highlighter pens and a refillable water bottle.

Preserved D9009 Alycidon at Crewe Gresty Bridge Open Day on the 23rd of July 2016. This was my last Deltic needed when I first saw it resting on the buffers at Kings Cross on the 10th of September 1977.

The locos on display had clearly been positioned with photographers in mind; I couldn't believe how grumpy some people became when their perfect shot was blocked by other enthusiasts wandering into their viewfinder. At one point, a dad was taking a carefully planned shot of his son in front of the Deltic, only to be met with lots of moans and grumbles from assembled camera-holding enthusiasts. It was completely unnecessary as there was plenty of time and room for everyone.

Before heading back home, I drove over to the electric depot on Wistaston Road, just south of the line from Crewe to Chester, however the access road was blocked by a security gate. With the depot out of view from the entrance, nothing could be seen, though a nearby trading estate provided a bit of a view of some of the stored 60s, 90s, 92s and a couple of rusty old shunters.

My final attempt to see anything that day involved driving around to try and get a view of the depot from across the other side of the Chester line. This area of land was previously part of the railway works but had been redeveloped when the size of the works was reduced to a much smaller footprint. Only a few remaining buildings of the once vast complex are now used for rail related activity. The owner, Bombardier, continues to employ about 1,000 people in component re-engineering. It is a pale shadow of the former site, which at its peak employed 20,000.

On the day of my visit in 2016, I found myself in the car park of McDonald's (along with a few others), trying to get a view across the track to the locos stabled outside the front of the electric depot. I saw a couple of 67s and a 66, but that was it.

To those queuing for the drive-thru, this must have seemed a strange activity, and I did note a few raised eyebrows and some tutting – especially when one of the enthusiasts went back to his car and returned with a small step ladder to get a better view. I cannot recall ever using a set of steps for such a purpose, and I could not make up my mind whether it was a bit weird or someone being creative.

As we stood there, I got chatting to a distinguished-looking old chap who was probably in his late sixties and was outfitted by Marks and Spencer. That sort of country gent look, lots of green check.

"It's so frustrating," he said, "my last two class 67s are over there somewhere but I can't see them!"

I've got to be honest; I don't think he'd been trying very hard; there were only thirty in the class, and they had been around for the last fifteen or so years. Maybe he was new at this?

After a final hopeful look, I went back to the car, re-joined the M6 and headed back to Bristol.

The Future Is Bright

Trainspotting and related activity is changing, some would say dying but I think that is a little harsh. Theorists on the subject, if such people exist, could probably talk for days on the reasons why this has happened and I'm sure there is some strong evidence to support its 'death'.

Though many consider the end of steam as being the most significant event in the trainspotting story, the big change happened almost unnoticed towards the end of the last century - the demise of the loco-hauled train and the contraction of the country's manufacturing/industrial base.

The reality; there were simply fewer engines left to spot - just a few hundred standard freight and mixed traffic locos – so the impact on the railway enthusiast's world was huge.

In the early 1980s, large numbers of engines were taken out of service with many making their way to Swindon for removal of any usable parts, prior to scrapping. At the Open Day on the 6th of June 1981, there were over a hundred locos littered around the site, including this long line of shunters stabled between the works and the main line.

A few years ago, I was on my way back from Swansea with a couple of hours to kill, so I turned off the M4 and drove into Newport, parking in the station car park adjacent to the loco stabling point. The loco yard was a ramshackle collection of portacabins and huts despite a new sign at the entrance proclaiming 'EWS Godfrey Road Locomotive Holding Yard' in the bright corporate EWS colour scheme.

Newport station, as always, was busy. Apart from the units, you can usually see a steady procession of freight rumbling through the station, and one of the regular traffic flows has been the Port Talbot to Llanwern ore and steel trains, which have been hauled by a variety of traction over the last thirty years. This includes double-headed 25s, triple-headed 37s, double-headed 56s, 60s and lastly, 66s. You can easily gauge how long you've been an enthusiast by the types of locos you've seen hauling these trains.

Newport is still a good place for freight. Here is 60040 on the 05:00 Robeston to Westerleigh. Taken on the 10th of March 2022.

On my impromptu visit, I counted about thirty people around the station who appeared to be rail enthusiasts of some sort. A variety of equipment was on show, ranging from the humble notebook to cameras, camcorders and handheld tape recorders. Yet again it struck me that nobody was under thirty years of age, and most were in their forties or fifties. Where were the kids?

I am sure the growth of computer games has reduced the appeal of hobbies that used to attract young people. Traditional hobbies that I grew up with such as stamp and coin collecting, bird watching, playing with toy soldiers and countless others have all been largely ignored or, perhaps, at the most, toyed with for a short period and then discarded.

So, whilst there is an argument to say that trainspotting is dying out as a result of the ending of the steam age, there is perhaps an even better argument for blaming it on the reduced number of locos. Yet I don't think either tells the full story. The bottom line is that technology has dampened a continued interest from the younger generation meaning that the humble trainspotter is under threat and may definitely be wiped out in the next twenty years. For a hobby that is so fulfilling as well as being packed with learning and opportunities, the potential 'death of the trainspotter' seems to me to be a crying shame.

If that is not pause for thought enough, consider the loss to railway stations of the sometimes slightly eccentric characters who populated the ends of platforms and often brightened everyone's day - and then there is the utilisation of the rail network itself. There is generally no longer any need to travel to certain locations to spot the few remaining engines, because most locos now move widely around the network. In the 'old days', engines tended to stay in certain areas, yet now, the class 66 working a freight out of Avonmouth today may be in Fort William tomorrow and at Toton the next day, for example. The plus side of this for those less inclined to travel in any case, is a reasonable chance, depending where you live, of clearing classes such as the 66s without leaving your home town.

Another major development in the trainspotting world has been the (necessary) tightening up of health and safety legislation, which means that permits for the few engine sheds which remain in existence are rarely issued, and open days are reduced to one or two every other year. Many depots are, in addition, also surrounded by high palisade fencing and can only be accessed through gates which are controlled by security codes. Perhaps that's a good thing, although the reality is that this increased security is more about protecting the rail operators from negligence claims than it is about any real concern from unwanted visitors.

Though we should not live in the past, sentimentality and nostalgia are synonymous with this hobby. Today's rail scene may arguably be less interesting than that of twenty, fifty or sixty years ago, but it is still important. If you look hard enough, you will still see the odd enthusiast recording and photographing some form of motive power. At the end of the day, what's the harm in that?

Open days, when they happen, are always well attended and it's great to see some of the rail operators embracing the interest shown in their industry. If local charities are also able to benefit from these events, then everyone's a winner. In Britain, we have such a rich railway heritage which creates and maintains its own brand. I, for one, firmly believe it should be celebrated.

The decision by First to relaunch the GWR (Great Western Railway) brand was a fantastic tribute to a proud legacy. The open day at St Phillips Marsh in Bristol in 2016 (to celebrate forty years of the high-speed train) was marked by one of the power cars from the first production unit being repainted in the 1976 blue/yellow livery. It was then named at the event in recognition of its designer, Sir Kenneth Grange, with the esteemed gentleman doing the honours.

The prototype High Speed Train on a test train at Bristol Temple Meads on the 20th of February 1975. Little did we know then that this design would be the mainstay of main line services in many parts of the country for the next forty years. Note the class 25 just trying to creep into the shot.

There was a good gathering of engines at the open day including a class fifty. Also, as a nod towards the location, an ex works condition shunter, 08663, was in attendance bearing the name of St Silas, a former local church in the area.

Preserved 50035 Ark Royal at St. Phillips Marsh Open Day on the 2nd of May 2016.

As you get older the past can draw you in. Recently, whilst in Cardiff, I took a spur-of-the-moment decision to visit Canton depot to see how it had fared over the last few years. As I walked along Ninian Park Road heading towards the depot, it struck me that this road, which I had walked along for the first time over forty years earlier, hadn't really changed. It was still quite scruffy with some infill where older properties had been demolished, but essentially, it was the same. The change in time was mainly evidenced by the cars now parked along it, these being very different to the Escorts, Marinas and Avengers of the 1970s.

As I was about to turn left into the small cul-de-sac that suddenly gave you a view across the front of Canton shed yard, I immediately slipped back in time and closed my eyes, hoping that when I opened them, I would see a panorama of 37s, 47s, 25s and 08s in front of me. Obviously, it was not to be. Instead, there were two orange and yellow Colas class 70s stabled outside the old, down-at-heel-looking running shed.

The main large cathedral-like maintenance shed is now used by Pullman Rail for component work, specialising in wheelset overhaul. The famous footbridge that had previously led from the end of the cul-de-sac to the depot had also gone; however, lurking at the back of the running shed was a shunter, 08499, which was a real throwback to an earlier age.

08663 St. Silas at Bristol ~ St. Phillips Marsh Open Day on the 2nd of May 2016. Named after a local church in the area, the shunter is now owned by the Avon Valley Railway Trust.

Constructed at Doncaster in 1958, it spent most of its life in Yorkshire - in fact, I first saw it stabled in the yard at Holbeck shed on a cold Saturday afternoon in April 1978. It had been reallocated to Canton in 2001 and was now owned by Colas. As I walked away from Canton, I did so with some happiness in my heart, knowing that the depot had at least survived.

Of course, technology has improved our lives dramatically over the last twenty years, and rail enthusiasts have not been ignored in this respect. It is now possible, for a small fee, to subscribe to railcam.uk, which allows you to watch trains pass by many strategically placed rail cameras dotted around the system; you don't even need to leave your armchair.

In fact, the company behind the scheme made play of the COVID pandemic by providing enthusiasts with the chance to watch trains whilst stuck at home in lockdown. However, if you are inclined to venture out of the house, there are also websites (e.g. *realtimetrains.co.uk*) that will tell you what trains are scheduled to pass at numerous locations across the network. The software pulls together details of various working timetables and even provides updates as to when the service passes a particular point and whether it is on time or late. In some cases, it will even tell you the number of the engine, which may explain to some degree why there aren't so many spotters waiting patiently on the end of platforms anymore; our hobby has become intelligence-driven, and there is no longer the need to leave things to chance.

I once idled away a pleasant couple of hours seeing how many locos I could spot using Google Street view. I managed to see about twenty, with my favourite being an ex-EWS class 66 working through Rouen. I don't know how often the street view is updated, but when I last looked, you could make out the loco numbers at Hoo Junction, Newport Alexandria Dock Yard, Bescot (from the M6), Peterborough and March.

I have to be honest and admit that I've never had a huge amount of interest in the preserved railway world; it's always been about the current scene for me; however, in 2019, I was given a Christmas present from Tracey for a diesel driver experience on the Bodmin and Wenford Railway. Though it was delayed (like everything else) due to COVID, it eventually took place on an overcast Sunday in October 2021 when I found myself clambering into the driver's seat of 50042. The morning was spent with just the engine running on its own to Boscarne Junction and back, and then in the afternoon, we hauled a few coaches on a round trip to Bodmin Parkway.

As an owner of two rather quick motorcycles (I have added a 900cc Yamaha alongside the Ducati), it's fair to say that I like travelling at speed, but sat at the controls of the class 50, I was more than a little nervous, even though there is a 25mph speed restriction on the line. As we headed back from Bodmin Parkway, my very patient teacher told me to go to 'full power' as we passed underneath the A38 road bridge, a short distance from Bodmin Parkway. The resulting noise was spine-tingling and something I will never forget, even if we were travelling at less than 30mph!

To end the day, we had a good look around the engine shed at Bodmin, where 37142 was having a major overhaul. I was massively impressed by the whole experience and the dedication shown by the people running the railway. Despite my earlier comment about my passion being purely of the 'now', I genuinely believe that I will continue to develop my interest in this area, too.

Me pictured with 50042 on a Diesel Driver Experience at the Bodmin Railway on the 9th of October 2021.

Despite all of these changes to my beloved hobby over the years, the basic attraction of watching a train pass by remains timeless. In 2017 Tracey and I moved into a bungalow in Pill, a quiet village between Bristol and Portishead, which is the closest I have ever lived to a functioning railway line. Pill is on the freight branch from Bristol to Portbury Dock, which uses the former Bristol to Portishead branch that closed to passengers in 1964. It eventually closed to all traffic in the early 1980's after which it lay rusting for several years until reopening for freight traffic in 2002.

It only runs as far as Pill, though, where a new branch line was laid into the docks to serve, primarily, a coal import facility. The subsequent demise of coal-fired power stations means that all this traffic has now gone, with the line only being used occasionally for ad-hoc traffic.

One of the other regular but now-ended traffic flows was a daily train taking imported cars up to Warrington and Mossend, just outside of Glasgow. Talking to one of the Pill residents recently, he said that local lads used to jump onto the car train as it slowed through Pill station and jump off again at Bedminster station, where it was held whilst waiting for a path into Temple Meads. Unfortunately there was one occasion when the car train wasn't held at Bedminster, and one lad was stuck on the train all the way to Mossend, where he was intercepted by Transport Police.

The closed station at Pill is a five-minute walk from our home, and one autumnal evening last October, I was walking back from the local supermarket by the former station when I heard a train approaching. I stopped in my tracks. At this location, the line is in a stone cutting as it approaches the old station, and although class 66s are very quiet compared to their predecessors, the engine could clearly be heard approaching. Up ahead of me was a dad walking with his son, who looked about four or five. As soon as he heard the train coming, he got excited, and as I watched, the dad lifted his little lad up so that he could see the train pass by.

A few other people stopped what they were doing and also looked over the bridge, all of us watching the 66 as it rumbled by with its long haul of empty wagons before disappearing into the dusk once more as it headed for Portbury Dock. Everything went quiet for a moment after that, and then people continued with their business. But in that one moment, I felt hugely reassured.

No matter what the future holds, no matter how far technology takes us, the magic of watching a train pass by on those parallel steel rails is something that will never go away.

Bibliography

B R Diesel and Electric Locomotive Directory	*Marsden C J, Oxford Publishing Company 1991*
BREL Locomotive Works	*Vaughan J, Oxford Publishing Company 1981*
Escort Carrier 1941-1945	*Poolman K, Ian Allan, 1972*
Great Western Engine Sheds 1837-1947	*Lyons E and Mountford E, Oxford Publishing Company 1979*
On Shed: A Survey of British Rail Depots Around The Regions	*Fisher A, Kelsey Media 2018 (a series of magazines)*
Shed by Shed, Part 7	*Walmsley T, St Petroc Info Publishing 2011*
Shed by Shed, Part 8	*Walmsley T, St Petroc Info Publishing 2012*
Shunter Duties	*Woodley G, Wood R, Inter City Railway Society Publication 1979*

www.ingramcontent.com/pod-product-compliance
Lightning Source LLC
Chambersburg PA
CBHW061125070526
44584CB00033B/4226